ExoticsCon 2021

AA006135

Building Exotics Excellence. One City.
One Conference.

Nashville, Tennessee, USA
28 August – 2 September 2021

ISBN: 978-1-7138-5432-6

Printed from e-media with permission by:

Curran Associates, Inc.
57 Morehouse Lane
Red Hook, NY 12571

Some format issues inherent in the e-media version may also appear in this print version.

Copyright© (2021) by Association of Reptilian & Amphibian Veterinarians
All rights reserved.

Printed with permission by Curran Associates, Inc. (2022)

For permission requests, please contact Association of Reptilian & Amphibian Veterinarians at the address below.

Association of Reptilian & Amphibian Veterinarians
5003 SW 41st Blvd
Gainesville FL 32608-4930
USA

Phone: 480-703-4941

info@arav.org

Additional copies of this publication are available from:

Curran Associates, Inc.
57 Morehouse Lane
Red Hook, NY 12571 USA
Phone: 845-758-0400
Fax: 845-758-2633
Email: curran@proceedings.com
Web: www.proceedings.com

TABLE OF CONTENTS

Feeling Blue?...Medicine and Surgery of Tiliqua Species

Dr Shane Simpson
The Unusual Pet Vets
210 Karingal Drive, Frankston, Victoria, 3199, Australia
E: shane@unusualpetvets.com.au
W: www.unusualpetvets.com.au

Blue-tongue lizards (*Tiliqua sp.*) are one of the most common lizard species kept as pets in Australia and their appeal has started to reach other parts of the world where access to these lizards is possible. In Australia their popularity lies in part to the fact that for several of the species available, a wildlife license is not required to keep them (unlike most native Australian reptiles being kept in captivity). In addition to this, selective breeding has led to the creation of some particularly stunning colour morphs that are only adding to their increasing popularity as pets.

In their native Australia they are the most common species of wild reptile that is brought into wildlife carers and veterinarians for treatment. Their ability to adapt to human surrounds unfortunately often places them in harms' way.

Species History and Taxonomy

The genus name, *Tiliqua*, was devised in 1870 and its origins are apparently meaningless. They are the biggest members of the skink (Scincidae) family and because of their unique tongue they are commonly known as blue-tongue lizards, blue-tongue skinks, blue-tongues or simply, blueys. The genus contains seven members (though as with most taxonomic descriptions there is some debate) that are found throughout Australia, Indonesia and New Guinea.

These species are:

1. Pygmy Blue-tongued Lizard (*Tiliqua adelaidensis*)
   This highly endangered species was thought to be extinct for over 30 years until in 1992 a researcher found the remains of an adult male in the stomach contents of a roadkill Brown Snake. There are now thought to be 5000-7000 individuals living in 31 small, isolated sites located on private agricultural land in south-eastern South Australia. There are only a small number of animals held in zoological collections and they are never likely to be introduced to the private market.
2. Centralian Blue-tongued Lizard (*Tiliqua multifasciata*)
3. Western Blue-tongued Lizard (*Tiliqua occipitalis*)
4. Blotched Blue-tongued Lizard (*Tiliqua nigrolutea*)
   It is generally regarded that there are two geographical forms of this species known as the lowlands and highlands (or alpine). There are also now several morphs being produced such as super pinstriped and axanthic.
5. Eastern Blue-tongued Lizard (*Tiliqua scincoides*)
   This is the most common species kept as pets in Australia. There are 3 subspecies:
   i. Eastern Blue-tongued Lizard (*Tiliqua scincoides scincoides*)

This subspecies is the one that breeders have really been busy creating all sorts of new morphs. This list grows almost daily but now includes T- Albino, T+ Albino, Hypermelanistic, Anery, Genetic Reduced Pattern, Lava, Snow, Platinum, Alabaster, Sunglow, Blackout, T+ White, Moonglow and Mercury.

ii. Northern Blue-tongued Lizard (*Tiliqua scincoides intermedia*)
These are the largest and heaviest of the species.

iii. Tanimbar Blue-tongue Lizard (*Tiliqua scincoides chimaerea*)
This species comes from Tanimbar Island, Indonesia. They are known to be the most aggressive of the blue-tongue species.

6. Shingleback (*Tiliqua rugosa*)
   The Shingleback is known by many names including Bobtail, Pinecone Lizard, Sleepy Lizard and Stumpy Tailed Lizard. There are 4 subspecies:
   i. Eastern Shingleback (*Tiliqua rugosa asper*)
   ii. Western Shingleback (*Tiliqua rugosa rugosa*)
   iii. Rottnest Island Shingleback (*Tiliqua rugosa konowi*)
   iv. Shark Bay Shingleback (*Tiliqua rugosa palarra*)

   At the 2018 ARAV conference in Atlanta, Georgia, I gave a presentation entitled "Armour Plated Medicine… Tips, Tricks and Treatments of Shingleback Lizards that provides specific advice on the veterinary care of this species.

7. Indonesian Blue-tongued Lizard (*Tiliqua gigas*)
   This species is endemic to New Guinea and various surrounding islands. There are several subspecies that are recognised:
   i. Indonesian Blue-tongued Lizard (*Tiliqua gigas gigas*)
      Three types of these are known:
      a. Classic
      b. Halmahera
      c. Irian Jaya
   ii. Merauke Blue-tongued Lizard (*Tiliqua gigas evanescens*)
   iii. Kei Island Blue-tongued Lizard (*Tiliqua gigas keiensis*)

Blue-tongue Lizard Facts

- Size ranges from 90mm SVL up to 380mm SVL
- Average weight is 200-400 grams
- Viviparous
- Juveniles are duller than the adults until they have their first shed and the banding on the body has greater contrast than that seen in adults. Juveniles also have a much darker blue tongue.
- Sexing is not easy and there is much misinformation available. Visual sexing may be possible when the lizards are at least 80% of adult SVL but there is considerable variation amongst individuals. Females are generally longer in the trunk than males with the heads and shoulders being proportionally smaller. Adult males may have obvious hemipenal bulges. Probing is not reliable in these species and hemipenal popping is difficult due to their size and strength. Hemipenal transillumination is not of use due to the heavy pigmentation and

scalation of the tail. The presence of hemipenal plugs can indicate a male but are not reliably produced by every male so their absence should not be used to assess an animal as female. Ultrasound sexing is very difficult due to the heavy scalation and resultant poor image quality. Mating behaviour is a good indicator of sex. The preferred method used by the author is that of radiographic dye. 0.5-1ml of iodinated dye is instilled using a crop feeding needle into the hemipenal area and a radiograph taken. A male will have a short, curled dye column while the female will have a long, tapering dye column.

- Tail autonomy is seen in all species other than Shinglebacks.
- Omnivorous. Blue-tongue lizards can be fed on fruits and vegetables supplemented with slow moving insects such as cooled crickets (to slow them down so as they can be caught) and mealworms, boiled egg and commercial lizard foods. One of their favourite foods are snails.
- Calcium and multivitamin powders should be used weekly in adults. Adults can be fed 2-3 times a week. Juveniles can be fed 3-4 times a week.
- Lifespan: 15-30 years. Shinglebacks are known to exceed 35 years.

Common Medical Conditions of Blue-Tongue Lizards

There are several specific medical conditions that afflict blue-tongue lizards.

Metabolic Bone Disease

MBD is term used to describe several syndromes seen in reptiles. By far the most common of these syndromes is Nutritional Secondary Hyperparathyroidism or sometimes referred to as Nutritional Metabolic Bone Disease (NMBD).

The symptoms of NMBD in blue-tongue lizards include:

i. Scoliosis and kyphosis (i.e., bending of the spine). This is thought to be due to their strong epaxial muscles pulling the soft spinal column into the unusual shapes.
ii. Trouble breathing if the scoliosis affects the ability of the lungs to expand or causes compression of the trachea.
iii. Lethargy.
iv. Paralysis of the back legs.
v. Tremors and spasms of the legs.
vi. Death

NMBD can be diagnosed by:

i. History of inadequate exposure to UVB light or inadequate calcium in the diet or both.
ii. Clinical signs.
iii. Radiographs. These will show a lack of bone density throughout the body.

Treatment of NMBD involves:

i. Provision of adequate UVB light. Exposure to unfiltered sunlight is the best source of UVB.
ii. If tremors or spasms are present, then an injectable calcium gluconate can be given and repeated every 6 to 12 hours until symptoms cease which usually takes 12 to 24 hours if oral calcium supplementation is commenced at the same time.
iii. Commence oral calcium supplementation.

iv. Calcitonin. It is given via injection for two injections 1 week apart. It is vital that blood calcium levels be normalised with at least 1 week of oral calcium supplementation before this is given as it has the potential of causing life-threatening hypocalcaemia. While no controlled scientific studies have been done in blue-tongue lizards, calcitonin has been seen to speed up the improvement in Green Iguanas with NMBD.
v. Stabilisation of fractures using careful external splinting.
vi. Correction of any husbandry deficits that led to formation of the problem.

Nidovirus

In 2016 a novel nidovirus associated with signs of respiratory disease was isolated from Shingleback Lizards in Perth, Western Australia. Commonly referred to a "Bobtail 'Flu", the virus was formally named Shingleback Nidovirus 1. The virus was subsequently found in Western Blue-tongue lizards, Centralian Blue-tongue lizards and Eastern Blue-tongue Lizards across most Australian states.

Affected animals show the following clinical signs:

i. Lethargy
ii. Anorexia
iii. Weight loss
iv. Sneezing
v. Mucopurulent nasal and ocular discharge

There is no specific treatment for the infection. Treatment is based on supportive care using antibiotics to treat secondary bacterial infections, administration of fluids (orally and subcutaneously) and assist feeding.

Dysecdysis

Like most lizards blue-tongues shed it in multiple pieces. This can create issues when they retain shed around their toes. This skin desiccates, constricts and occludes the blood supply. The toes become necrotic and slough. In some cases, the dead toes may need to be surgically removed.

Hemipenal Abscess

Like other lizard species, male Blue-Tongue lizards have two hemipenes that sit within pockets in the cranioventral part of the tail. These pockets can sometimes become infected and fill with pus. Due to the fibrous nature of the material antibiotics alone will not treat the issue and surgical debridement is required.

With the animal positioned in lateral recumbancy the pus can be manually expressed. A probe can then be inserted into the affected hemipenal pocket. This creates a guide for the incision to be made to open up the pocket. A section of skin over the cavity can be removed to allow access to the pus. The soft tissue edges of the wound can be sutured to the surrounding skin to marsupialise the wound and keep it open. Post-operatively the owner can continue to flush the pocket with antiseptics (e.g. chlorhexidine) and apply topical antibiotic cream such silver sulfadiazine. Analgesia and systemic antibiotics are also used. The lizard should be housed on newspaper to avoid substrate material from getting into the wound during the healing process.

The author has never seen a hemipenal abscess in a wild blue-tongue lizard and so it would appear this is a condition of captivity.

Ticks

Wild Blue-tongue lizards are often presented to veterinary clinics in Australia due to tick infestation These "bush ticks" are of assorted species and are commonly found in the ear canals and axilla of the lizards. They do not appear to cause any great problems, but heavy infestations could theoretically result in significant blood loss. The author has seen one lizard showing ataxia until the ticks were removed from the ears and the lizard was immediately normal. Ticks in the ear canals will often result in large amounts of skin being retained in the canals.
Treatment involves simply removing the ticks with a pair of forceps. They can be treated with a pyrethrin-based product and allowed to die before removal.

Mites

The most common parasitic problem suffered by Blue-tongue lizards is infestation with reptile mites and the most common of these is *Ophionyssus natricis*. These small blood-sucking mites can be very difficult to eradicate once established in an enclosure.
Symptoms of mite infections include visibly seeing the mites, thickened and retained skin and lizard that spends much of its time sitting in its water bowl. The most telling symptom is the appearance of a white, dusty powder over the body of the lizard.
There are many and varied treatments that have been tried to control reptile mites. No matter what is used it is important to understand that both the animal and the enclosure must be treated. Using fipronil wiped onto animal and a pyrethrin-based spray in the enclosure is effective method of eradicating the infestation. Products available for treating mites will be dependent on what is available in various countries.

Gout

Gout is a common condition seen in many species of reptiles and blue-tongue lizards are no exception. In *Tiliqua sp.* the elbow joints seem to have a predilection for accumulating urate. Affected animals will present non-weight bearing on their front legs with obviously swollen elbow joints. Aspirates and cytology of the joints reveal excessive urate crystals. In severe cases the joints may need to be flushed to remove as much of the urate as possible. Blood sampling can be performed to assess the blood uric acid level. Treatment consists of improving hydration, reducing animal proteins in the diet, the use of allopurinol to reduce urate production and providing analgesia.

Trauma

Blue-tongue lizards are very well adapted to living around humans and as such encounter dogs, lawn mowers and brush cutters quite often.
Injuries that will not allow the animal to function normally when released should preclude that it be euthanased. Special attention should be paid to penetrating wounds to the body cavity. Fractures to the spine and long bones should be assessed with radiographs.

Corneal Ulcers

Due to their terrestrial lifestyle and tendency to use their heads to push through substrate, blue-tongue lizards are often presented for corneal damage. Affected animals will show blephrospasm and sometimes a mild serous ocular discharge from the affected eye. The use of fluorescein can be used to visualise the corneal ulcer. Treatment with topical antibiotic eye drops and oral anti-inflammatories such as meloxicam can be used. Treatment for upwards of 3 weeks is often needed until complete resolution is achieved.

Keratoconus

A condition that is seen occasionally in this species appears to be very similar to one known as keratoconus in humans. This condition is characterised by a thinning of the central cornea such that it begins to bulge out in a dome shape. In the early stages the animal seems unaffected but with time exposure keratitis can develop and the cornea can become so traumatised that enucleation of the eye is required. The cause is unknown. Treatment with topical artificial tears and antibiotic preparations can slow the progression.

Diarrhoea

A common presentation seen in captive *Tiliqua sp.* is diarrhoea. Affected animals will be generally bright and alert on clinical examination. They will be in good body condition and eating well. They produce profuse, foul-smelling diarrhoea.

Faecal smears often reveal excessively large numbers of flagellates.

Treatment with oral metronidazole at 10mg/kg for 10 to 14 days rectifies the problem.

Dystocia

Dystocia is rare in blue-tongue lizards. Causes include abnormally large foetuses and pelvic deformities preventing passage of the foetuses. Treatment involves performing a caesarean to remove the retained foetuses.

Neoplasia

In the author's experience neoplasia of any type is rare in *Tiliqua sp.*

Viruses of Reptiles in Clinical Practice

Rachel E. Marschang, PD Dr med vet, DECZM (Herpetology), FTÄ Mikrobiologie

From Laboklin GmbH & Co. KG, Steubenstr. 4, 97688 Bad Kissingen, Germany

Introduction

A wide range of viruses have been described in reptiles, with the number of reports of newly discovered viruses increasing rapidly. The clinical importance of many is not yet known. At the same time, diagnostic testing for viruses in reptiles is also improving, and keeping up with the latest developments in what to test for, how to test for it, and how to interpret the results you get back from the lab can be difficult. This presentation will provide an overview of select viruses found in reptiles and then concentrate on those associated with specific disease processes and their diagnosis. Particular focus will be on viruses in the families *Adenoviridae*, *Herpesviridae*, *Iridoviridae* (ranaviruses and invertebrate iridoviruses), *Tobaniviridae* (order *Nidovirales*), *Arenaviridae* (the causes of inclusion body disease in boas and pythons), *Paramyxoviridae* (ferlaviruses), and *Picornaviridae*. A more detailed description of viruses described in reptiles can be found in Marschang et al.[1] An overview of the viruses discussed in this presentation and their diagnosis can be found in Tables 1 and 2.

In addition to the developments in virus discovery, significant effort has been made in recent years to better understand the connection between infection and disease for specific viruses in reptiles. In many cases this relationship appears to be complicated. The development of disease can depend on many different factors. These can include the host species and age; the virus species and strain; the season and environmental temperatures; stress, hygiene, the presence of other pathogens, and likely many others. In many cases, viruses have been shown to persist within their reptile hosts for extended periods of time, with or without clinical signs of disease. Determining freedom of infection can be quite difficult or impossible in some cases, and risk assessments for introduction of infection into a collection as well as the development of quarantine protocols should consider possible outcomes of infection, options for virus detection, and infection status within the existing collection as well as including long-term plans for what to do with animals that test positive for a virus but appear clinically healthy.

Viruses of chelonians

A wide variety of viruses from at least 18 different families have been described in the Testudines. The viruses most often described in chelonians presented in clinical practice are herpesviruses. In pet chelonians, herpesviruses have been best studied in tortoises, in which four different types (testudinid herpesvirus (TeHV) 1-4) have been characterized so far. These viruses are generally associated with upper digestive and upper respiratory tract disease, often with the formation of diphtheroid-necrotizing plaques on the tongue of affected susceptible tortoises. Herpesviruses have also been described in a variety of turtle species, in which genetically diverse viruses have been found. All of the herpesviruses found in chelonians so far do, however, appear to cluster together with members of the genus *Scutavirus* in the subfamily *Alphaherpesvirinae*. Other viruses regularly described in chelonians presented to veterinarians in practice include ranaviruses, adenoviruses, and picornaviruses (Table 1).

Viruses of squamates

Viruses from at least 25 different families have been described in squamate reptiles. In general, virus infections in snakes have been better studied than those in other groups of squamates, and more is known about diagnosis, disease, and prevalence for several of these.

Arenaviruses are the cause of inclusion body disease (IBD) in boas and pythons. Prevalence in captive boas has been shown to be very high, with many infected snakes remaining clinically inapparent for extended periods of time. Diagnosis in pythons is more difficult and several studies have shown that both inclusion bodies and virus in infected pythons can be restricted to the brain, making diagnosis in live pythons challenging or impossible.

Viruses in the order *Nidovirales*, family *Tobaniviridae*, subfamily *Serpentovirinae* have been shown to cause severe respiratory disease in pythons with a high prevalence of infection reported in captive pythons in the USA and Europe. Related but genetically distinct viruses have also been found in snakes in other families, as well as several lizard species and a turtle species.

Other viruses that have been described repeatedly in captive snakes include adenoviruses and reoviruses. Adenoviruses have also been described in a variety of lizard species. The vast majority of these have been in the genus *Atadenovirus*. The species *Lizard atadenovirus B* includes the bearded dragon adenovirus 1, also known as agamid atadenovirus 1, which has been found in captive bearded dragons (*Pogona vitticeps*) in many different countries around the world, in some cases with high prevalence. The clinical significance of this virus in these animals is not yet well understood. In many cases, co-infections with a variety of other possible pathogens have been described, and it is hypothesized that this adenovirus may be involved in multi-factorial disease processes. The host age may also play a role in disease development. Reoviruses have also been described regularly in various lizard species. Like adenoviruses, they may be involved in multifactorial disease processes. Herpesviruses have been described in squamate reptiles in a number of cases, both in clinically healthy animals as well as in animals with a variety of lesions affecting, among others, the skin, oral mucosa, liver, spleen, intestine, and venom glands in elapid and viperid snakes.

Diagnosis of viral infections in reptiles

Virus diagnostics in reptiles is a rapidly developing field. In general, it can rely either on direct detection of the virus or parts thereof (e.g. viral DNA or RNA, viral proteins) or on detection of a host immune response to the virus.

Direct virus detection is most often used for the diagnosis of viral infection in reptiles. This requires the use of appropriate samples and an understanding of the biology of the virus. Virus isolation, PCR, immunohistology and in situ hybridisation have all been used to detect viruses in samples or tissues from reptiles. More recently, next generation sequencing (NGS) methods have been used to detect viral genome sequences in samples from reptiles. The most common method used for the detection of specific viruses in reptiles is the PCR. It is, however, important to understand what exactly the method used is able to detect and how results are verified.

Detection of an immune response generally relies on antibody detection methods. Virus neutralization, hemagglutination inhibition, and ELISAs have all been described for detection of antibodies against specific viruses in plasma or serum of reptiles. The reptile immune response is dependent on multiple factors, complicating the use of these methods. They are mostly of interest for quarantine examinations to determine possible infection with viruses known to cause persistent or latent infections, e.g. herpesviruses or for determining the presence and prevalence of specific viruses in a population. However, it is important to understand what viruses, specifically what serotypes, are being used in the test.

When contemplating virus diagnostic testing, it is a good idea to contact your laboratory to discuss testing options and interpretation. Any virus diagnostic testing should always be interpreted in conjunction with additional information including e.g. clinical examination, hematology, blood biochemistry, cytology, and histopathology.

Key words: adenovirus, herpesvirus, nidovirus, reptarenavirus

Disclosure statement: The author is employed by a commercial veterinary laboratory.

Literature

1. Marschang RE, Origgi FC, Stenglein MD, Hyndman TH, Wellehan JFX, Jacobson ER. Viruses and viral diseases of reptiles. In: Jacobson ER, Garner MM, eds. Infectious Diseases and Pathology of Reptiles Volume 1, Color Atlas and Text, 2nd Edition. Boca Raton, LA, USA: CRC Press; 2021: 575-703.

2. Marschang RE, Meddings JI, Ariel E. Viruses of Reptiles. In: Hurst CJ (ed.) Studies in Viral Ecology. Wiley. In press.

Additional literature can be found in these review chapters and is also available from the author.

Table 1: Select viruses commonly found in pet reptiles and their associated pathologies (adapted from Marschang et al.[2])

Virus family	Genus	Species	Hosts	Commonly described associated pathologies
DNA viruses				
Herpesviridae	*Scutavirus*	*Testudinid alphaherpesvirus 3*	Tortoises	Stomatitis, rhinitis, glossitis, esophagitis
	Other probable scutaviruses		Chelonians	Oral lesions, skin lesions, hepatitis, no clinical disease in some cases
	Not classified		Squamates	Oral lesions, skin lesions, hepatitis, no clinical disease in some cases
Adenoviridae	*Atadenovirus*	Several	Squamates, chelonians	Mostly gastrointestinal tract and liver affected.
	Siadenovirus		Chelonians	Severe systemic disease
	Testadenovirus		Chelonians	Changes in the liver reported in some animals. No disease detected in some cases
Iridoviridae	*Ranavirus*	*Frog virus 3*, *Common nidwife toad virus*, others	Chelonians, squamates	Changes in the liver, skin, spleen, and other tissues
	Iridovirus	*Invertebrate iridescent virus 6-*like	Squamates	No disease noted in many cases, in some cases skin lesions, cachexia
RNA viruses				
Reoviridae	*Orthoreovirus*	*Reptilian orthoreovirus*, *Testudine orthoreovirus*, others	Squamates, occasionally chelonians	Respiratory disease, neurologic disease, gastro-intestinal tract changes, hepatitis
Paramyxoviridae	*Ferlavirus*	*Reptilian ferlavirus*	Squamates, occasionally chelonians	Pneumonia, other tissues can also be affected
Arenaviridae	*Reptarenavirus*	*California reptarenavirus*, *Giessen reptarenavirus*, *Golden reptarenavirus*, *Ordinary reptarenavirus*,	Boas, pythons, rarely other snakes	Inclusion body disease (IBD), associated mostly with neurologic disease as well as changes in a wide variety of tissues, including

		Rotterdam reptarenavirus		respiratory, hepatic, gastro-intestinal, and dermatological disease
Picornaviridae	*Torchivirus*	*Torchivirus A*	Tortoises	Kidney disease, severe softening of the carapace in juvenile tortoises, rhinitis
Tobaniviridae (order *Nidovirales*, subfamily *Serpentovirinae*)	*Pregatovirus*	*Ball python nidovirus 1, Morelia tobanivirus 1,*	Snakes	Respiratory and upper digestive tract disease
		Shingleback nidovirus 1	*Tiliqua* spp.	"Bobtail flu": upper respiratory tract disease
		Berisnavirus 1	Turtles (Bellinger river turtle)	Ocular lesions, nasal discharge, severe disease outbreak

Table 2: Diagnosis of select viruses infecting reptiles (in alphabetical order)

Virus family	**Properties**	**Genus**	**Diagnostic test: virus detection**	**Samples for virus detection**			**Serology (detection of antibodies in plasma or serum)**
				Animals that appear clinically healthy	**Clinically ill animals**	**Dead animals**	
Adenoviridae	Non-enveloped, dsDNA	*Atadenovirus*, also *Siadenovirus*, *Testadenovirus* in chelonians	PCR	Cloacal swabs	Cloacal swabs	Liver, intestine	Virus neutralization described for some atadenovirus strains. Not commercially available
			Histology			Liver, intestine	
Arenaviridae	Enveloped, ssRNA	*Reptarenavirus*	PCR	Esophageal swabs, oral and cloacal swabs, whole blood, biopsies of: liver, esophageal tonsils	Esophageal swabs, oral and cloacal swabs, whole blood, biopsies of: liver, esophageal tonsils	Brain, liver, pancreas, kidney, other tissues	None available
			Histology/cytology	Whole blood, buffy coat biopsies of: liver, esophageal tonsils	Whole blood, buffy coat biopsies of: liver, esophageal tonsils	Brain, liver, pancreas, kidney, other tissues	
Herpesviridae	Enveloped, dsDNA	*Scutavirus* in chelonians	PCR	Oral swab	Oral swab, lesions	Lesions, tongue, liver, brain	Virus neutralization and ELISA described for some strains found in tortoises

Iridoviridae	Enveloped[1], dsDNA	*Iridovirus* (invertebrate Iridoviruses (IIV))	PCR	Not applicable	Skin biopsies	Skin, liver	None available
		Ranavirus	PCR	Oral and cloacal swabs, whole blood, tissue samples[2]	Oral and cloacal swabs, whole blood, tissue samples[2], lesions	Lesions, liver, tongue, intestine, skin, kidney	None available
Paramyxoviridae	Enveloped, ssRNA	*Ferlavirus* (aka ophidian paramyxovirus, OPMV)	PCR	Tracheal wash, oral and cloacal swabs	Tracheal wash, oral and cloacal swabs	Lung, intestine, pancreas, kidney, brain	Hemagglutination inhibition against various strains
Picornaviridae	Non-enveloped, ssRNA	*Torchivirus* (virus "X")	PCR, virus isolation	Oral swab	Oral swab, cloacal swab	Tongue, esophagus, intestine, trachea, kidne, other tissues	Virus neutralization. Not widely commercially available
Reoviridae	Non-enveloped, dsRNA	*Orthoreovirus*	PCR, virus isolation	Oral swab, cloacal swab	Oral swab, cloacal swab	Tongue, lung, skin, intestine, brain, liver	Virus neutralization described for some isolates. Not commercially available
Tobaniviridae (Order *Nidovirales*, Subfamily *Serpentovirinae*)	Enveloped, ssRNA	*Pregatovirus*	PCR	Oral swab, tracheal wash, cloacal swab	Oral swab, tracheal wash, cloacal swab	Lung, oral mucosa, other tissues	None available

[1]Some of these do not require their envelope to be infectious; [2]Tissue samples have been shown to be much more sensitive for ranavirus detection than oral or cloacal swabs.

Managing Cloacal Prolapses in Reptiles

Christoph Mans, Dr. med, vet., Dipl. ACZM, Dipl. ECZM (ZHM)

From the School of Veterinary Medicine, University of Wisconsin-Madison, 2015 Linden Drive, Madison, WI 53706, USA

INTRODUCTION

Cloacal prolapses are commonly diagnosed in reptiles, and in a retrospective study, the incidence of cloacal prolapses in reptile patients was 1.9%.[1] Prolapses were more frequent in lizards and chelonians than in snakes, and the most common prolapsed organs were the penis/hemipenis.[1] The gastrointestinal (GI), urinary and reproductive tract all terminate in the cloaca in reptiles and therefore, oviduct, urinary bladder, penis (turtles and tortoises), or hemipenes (lizards and snakes), and the intestine can prolapse through the cloacal opening. Furthermore, the cloaca itself may evert and prolapse (i.e., true cloacal prolapse)[1].

Correct identification of the prolapsed tissue can be challenging, but it is critical to select the most appropriate treatment approach. In particular, if the tissue had been prolapsed for several hours to days, then changes its appearance due to tissue necrosis can make identification challenging. The tissue should be carefully manipulated and cleaned to aid identification.

DIAGNOSTICS

<u>Whole-body radiographs</u> or computed tomography should be performed in most reptiles presenting with a cloacal prolapse. Radiographs will help rule out the presence of mineralized eggs, which may be associated with increased straining and potentially lead to cloacal prolapse. Radiographs can also rule out the presence of uroliths, which can be either located in the urinary bladder or have migrated from the bladder into the cloaca. Snakes and some lizard species (e.g., bearded dragons) do not have a urinary bladder, which should be considered before evaluating these species. In addition, the GI tract can be assessed by radiographs for the presence of impactions (e.g., ingesta, sand, etc.) or tympany. Poor bone quality is most likely indicative of nutritional secondary hyperparathyroidism, and the patient may be hypocalcemic, which may lead to cloacal prolapse. Additional imaging techniques such as computed tomography or computed tomography are very useful for evaluating coelomic organs and evaluating coelomic effusion (e.g., due to yolk coelomitis). Oral contrast studies, using barium or iohexol, can also be considered, if advanced imaging options are not available.

<u>Routine blood work</u> is indicated in cases where a metabolic or systemic underlying disease process is suspected. A plasma biochemistry panel will allow to rule out hypocalcemia, impaired renal function, amongst other abnormalities. Increased calcium levels in female reptiles indicate reproductive activity, and therefore the cloacal prolapse may be associated with reproductive tract activity or disease. A complete blood count (CBC) will allow to rule out anemia (due to acute blood loss, e.g., from the prolapsed cloacal tissue) or due to chronic disease) as well as help to evaluate for a systemic inflammatory response. An increase in the total white blood cell count and morphological changes in the granulocytes support the diagnosis of an inflammatory process. However, without baseline values for each individual patient obtained during a period

then no underlying disease process was present, it may be challenging to interpret a single CBC result obtained during a disease process.

Fecal parasitology should be performed in most reptiles diagnosed with cloacal prolapses, particularly in animals with intestinal prolapse. Heavy parasite burden can lead to enteritis, diarrhea, straining, and ultimately cloacal prolapse. Therefore, fecal samples should be evaluated as a wet-mount preparation as well as flotation. If a fecal sample is not readily available, then an enema may promote defecation. Consider empiric antiparasitic treatment if a fecal sample cannot be obtained, but endoparasites are a likely underlying cause.

Saline-infusion cloacoscopy under sedation or general anesthesia is the most valuable diagnostic tool to evaluate cloacal prolapse cases. For this procedure, a rigid endoscope housed in a working sheath is used. An IV infusion set is connected to a fluid bag and one of the ports of the working sheath. The infusion of the cloaca with 0.9 % saline solution or a balanced electrolyte solution will distend the cloaca and allow for evaluation of the normal cloacal structures and the origin of the cloacal prolapse, which allows for the identification of the origin and type of the prolapsed cloacal tissue by exclusion. If, for example, the urinary bladder and rectum both can be identified during saline infusion cloacoscopy, then it can be ruled out that these organs have prolapsed. Manipulation of the cloacal prolapse and following the prolapsed tissue to its base will help identify the prolapsed tissue's origin.

TREATMENT

Treatment of cloacal prolapses should only be performed once the tissue has been identified. If the tissue is viable, repositioning of the prolapse under sedation, intrathecal anesthesia or general anesthesia can be attempted if it has been determined that reposition is the preferred treatment method. Repositioning is performed by gentle pressure with cotton-tipped applicators after applying hyperosmotic solution or sugar on the tissue. Alternatively, if the prolapsed organ has a lumen (i.e., intestine, cloaca) can be infused with saline, which will aid in repositioning. Once repositioned, two sutures are placed at the side of the vent to reduce the chances of further prolapse. The use of purse-string suture, although widely used in the past, is not recommended due to the risk of damage to the cloacal sphincter.[2] In lizards and snakes, percutaneous tagging sutures can also be used if the cloaca itself is prolapsed. If these techniques fail to prevent the reoccurrence of cloacal tissue prolapse, then cloacopexy or colonopexy should be considered.[2]

If a hollow organ prolapse (e.g., colon, oviduct) has occurred and the tissue has suffered substantial trauma, resection and anastomosis should be performed. With the reptile anesthetized in dorsal recumbency, the prolapsed tissue is cleaned. A cylindrical stent (e.g., a syringe case) is placed in the lumen, and orthogonal needles transfix the tissue. The nonviable tissue is excised, and mattress sutures are placed all around the organ. With this approach, continuity of the hollow organ is maintained. In oviductal prolapse cases, substantial oviductal disease and/or coelomitis are usually present, and these cases often benefit from coeliotomy and a more thorough assessment. Oviductal prolapse usually results in trauma to the oviductal vasculature, which can lead to substantial hemorrhage and hemocoelom.

Besides identifying the type of prolapsed tissue and providing immediate care to prevent further tissue trauma, it is crucial to investigate the primary underlying cause of the cloacal prolapse. Without identifying and correcting the primary underlying cause, the prognosis for the resolution of most cloacal prolapses is poor. While the underlying or concurrent diseases are being diagnosed, it is important to prevent further trauma to the prolapsed cloacal tissue.

CLOACAL EVERSION

The cloaca itself may evert and prolapse without any other organ system being involved.[1] Causes for cloacal eversion (true cloacal prolapses) include metabolic causes, straining, increased intracoelomic pressure, obesity, dystocia, inflammation/infection of the cloaca, amongst others. The everted and prolapsed cloaca may be mistaken as the urinary bladder, but the urinary bladder wall is thinner and has no lumen. The prolapsed tissue should be replaced if viable as soon as possible and stay sutures placed to reduce the cloacal opening.

PENILE OR HEMIPENILE PROLAPSE

Prolapse of the penis (syn. phallus) in chelonians is common. It can occur secondary to behavioral reasons (e.g., fighting males, agitated males during breeding season) or due to underlying disease (e.g., shell deformation leading to increased coelomic pressure, neurological causes, trauma. Penile prolapse secondary to cystic or cloacal calculi has been reported, and therefore radiographs should always be performed to rule out this condition.[3] Penile trauma can cause severe and life-threatening hemorrhage, and consequently, the penis should be carefully examined for often very small traumatic lesions. In lizards and snakes, the hemipenes can prolapse due to irritation from probing or from the iatrogenic protrusion of the hemipenes performed for sexing. Other causes include smegma accumulation in inflammation of the hemipenes. In many cases the underlying cause remains unknown. The author recommends surgical amputation for all penile prolapses in chelonians due to the high risk of reoccurrence unless breeding ability needs to be maintained. In lizards and snakes, the hemipenes can sometimes be carefully replaced and stay sutures placed. However, surgical amputation is recommended if the hemipenes are enlarged, or the tissue is nonviable or very inflamed. Breeding ability can be maintained if only one hemipenis is amputated. Amputation may be performed in toto, as the phallus and hemipenes do not contain the urethra.[2] In small individuals, one to two transfixing ligatures are placed at the base of the penis/hemipenis. The tissue is excised distal to the ligatures. In larger chelonians, each half of the phallus may be individually double ligated with encircling absorbable sutures. The penile stump can be closed with a simple continuous or a purse-string suture, particularly in larger species.[2] Tortoises have two cartilaginous cords within the penis, making transection and ligation more challenging than turtles.

OVIDUCTAL PROLAPSE

Prolapse of the oviduct is frequently associated with reproductive activity and is most common in chelonians and lizards. The prolapsed oviduct often appears plicated, thin-walled, and has a lumen. However, with a prolonged duration of the prolapse, the tissue is usually swollen, necrotic, or covered by a fibrinous scab. Diagnostics should always include diagnostic imaging to assess the reproductive tract. Causes include reproductive disease (oviductal impaction, salpingitis, yolk-coelomitis, dystocia) or increased intracoelomic pressure. Resection of the prolapsed tissue at the level of the entry into the urodeum of the cloaca has been reported as a treatment option. Still, it does not address the underlying disease, and therefore a coeliotomy is usually necessary to treat the underlying condition, and an ovariosalpingectomy is frequently performed. Intracoelomic bleeding is a common complication of oviductal prolapses due to trauma of the oviductal vessels.[2]

COLON/RECTAL PROLAPSE

Prolapsed rectum or colon have a lumen and a thick wall compared to other prolapsed organs. Usually, substantial swelling is present as the prolapse persists. Causes include endoparasitism, straining, constipation, increased coelomic pressure, or metabolic disease (e.g., hypocalcemia). If viable, the tissue may be replaced. If necrosis is evident, resection and end-to-end anastomosis are necessary.

URINARY BLADDER PROLAPSE

Prolapsed urinary bladders have a very thin wall, which is easily traumatized. The prolapsed tissue has no lumen, and often the prolapsed tissue is of substantial size. Underlying causes are similar to other cloacal prolapses. Cystitis, urolithiasis, and increased coelomic pressure are common causes. Viable bladder tissue can be carefully replaced by infusing the organ with saline. If the tissue is necrotic, amputation should be performed. The ureters in reptiles do not enter into the urinary bladder but into the cloaca's urodeum; therefore, trauma or ligation of the ureters is not a concern during cystectomy. The procedure can be performed transcloacally or intracoelomically, in particular in lizards.

REFERENCES

1. Hedley J, Eatwell K. Cloacal prolapses in reptiles: a retrospective study of 56 cases. *J Small Anim Pract* 2014;55:265-268.
2. Di Girolamo N, Mans C. Reptile Soft Tissue Surgery. *Vet Clin North Am Exot Anim Pract* 2016;19:97-131.
3. Huang J, Eshar D, Andrews G, et al. Diagnostic Challenge. *Journal of Exotic Pet Medicine* 2014;23:418-420.

Contemporary Topics in Chelonian Medicine, Surgery, Rehabilitation, and Conservation

Charles Innis, VMD, Dipl ABVP(Reptile and Amphibian Practice)

Affiliation: From New England Aquarium, 1 Central Wharf, Boston, MA 02110, USA

Session #?

Abstract: Turtles and tortoises are globally threatened due to anthropogenic factors. Many ill or injured chelonians can be successfully managed medically and surgically. This presentation reviews several realms of chelonian medicine and conservation including observations on subcutaneous drug therapy, non-steroidal anti-inflammatory drugs, methods for inducing oviposition, pre-femoral surgery, decompression sickness, endocrine observations during rehabilitation and transport, methods for post-release monitoring of rehabilitated turtles, and the recently established AZA SAFE American Turtles Program.

Key words: turtle, tortoise, chelonian, rehabilitation, subcutaneous, ketoprofen, oxytocin, dexmedetomidine, prostaglandin, pre-femoral, corticosterone, aldosterone, transport, satellite tag, acoustic tag, PIT tag, decompression sickness, AZA SAFE

SUBCUTANEOUS DRUG THERAPY

A number of studies have demonstrated that subcutaneous injections are practical and clinically useful for chelonians, resulting in therapeutic plasma drug concentrations (eg, oxytetracycline, voriconazole, fluconazole, hydromorphone).[1,2] This route is attractive because several subcutaneous sites are readily accessible in most turtles. Subcutaneous injection is minimally invasive (eg, compared to intracoelomic injection), and it reduces muscle trauma (e.g., compared to IM injections). The pre-femoral and pre-scapular regions are often used, with consideration that pre-femoral drugs may be influenced by renal portal or hepatic first-pass elimination or metabolism, resulting in lesser effect.

NON-STEROIDAL ANTI-INFLAMMATORY DRUGS (NSAIDs)

Non-steroidal anti-inflammatory drugs are often used for chelonian analgesia. While the use of COX-2 selective drugs is common, observations on COX enzyme expression in turtles suggests that the use of non-selective drugs may be prudent.[3] Despite pharmacokinetic data, there is a lack of pharmacodynamic data for NSAIDs in turtles, and their analgesic efficacy is not well studied. Meloxicam pharmacokinetics vary by species, making extrapolation difficult. Meloxicam may be minimally useful for loggerhead turtles (*Caretta caretta*) due to low plasma concentrations, but useful for green turtles (*Chelonia mydas*) with a five day dosing interval, and useful for Kemp's ridley turtles (Lepidochelys kempii) with a twice daily dosing interval.[4] A series of studies has confirmed theoretically effective plasma levels and multi-dose safety for the non-selective drug, ketoprofen, in loggerhead turtles.[5] Ketorolac appears to have favorably pharmacokinetic properties for eastern box turtles (*Terrapene carolina carolina*) but not for loggerhead turtles.[6,7] The lesser known drug, tolfenamic acid appears to have favorable properties for long dosing intervals for red-eared slider turtles (*Trachemys scripta elegans*), and loggerhead turtles, and is worthy of increased clinical attention.[8,9]

INDUCING OVIPOSITION

Egg retention is a common clinical problem for turtles. Traditional interventions including calcium, fluid therapy, oxytocin, and provision of a nesting environment are often, but not always, effective. Alternative induction techniques have been developed. While there is no thorough peer reviewed study on the subject, clinicians should consider the use of prostaglandin F2α injections, and α2 agonists as adjuncts or alternatives to conventional therapy. Recent clinical cases indicate efficacy of such methods for eastern box turtles, a Ryuku black breasted leaf turtle (*Geoemyda japonica*), an elongated tortoise (*Indotestudo elongata*), and a Blanding's turtle (*Emydoidea blandingii*). Substantial unpublished data also supports the use of these drugs in many other chelonian species.[10] Of course, no method will be effective in every situation, and clinicians must consider the possibility of ectopic eggs or other disease states if medical induction fails repeatedly.

PRE-FEMORAL SURGERY

The importance of pre-femoral surgery in chelonians has been recognized for many years. Veterinarians should ALWAYS consider whether a problem can be addressed by pre-femoral surgery prior to embarking on more aggressive approaches (i.e., plastronotomy). Pre-femoral surgery may allow definitive management for disorders of the digestive, urinary, and reproductive system, and may allow for biopsy of most major viscera, including the lungs. Recent clinical cases, including management of a prolapsed oviduct in a Speke's hingeback tortoise (*Kinixys spekii*) and lung biopsy in a Kemp's ridley turtle, provide evidence of the utility and safety of this method.

DECOMPRESSION SICKNESS

A growing body of evidence indicates that sea turtles can be affected by decompression sickness when entrapped at depth and then rapidly brought to the surface.[11] Severe gas embolism has been noted in the brain, heart, liver, mesenteric vessels, and kidneys. Hyperbaric therapy may be effective if instituted quickly.[12]

ENDOCRINE OBSERVATIONS DURING REHABILITATION AND TRANSPORT

Plasma corticosterone concentrations may be very high when turtles are first admitted to rehabilitation, and often decline over time as turtles convalesce. During long-distance transport for release to the wild, corticosterone may increase substantially, along with other evidence of physiologic stress such as increased blood glucose, leukocytosis, heterophilia, and increased heterophil to lymphocyte ratio. Such effects can be mitigated by thoughtful planning to reduce stimuli and transport duration, and an overnight recovery period prior to release may also be useful. The impacts of elevated corticosterone are not fully known for turtles, but it is presumed, as for other species, that it may inhibit immune function and healing. Very recent unpublished data in the author's laboratory indicate that aldosterone concentrations may also be substantially elevated in turtles during periods of stress, which could theoretically affect blood electrolyte status.

POST-RELEASE MONITORING OF REHABILITATED TURTLES

Assessment of the post-release survival of rehabilitated wildlife is important for determining the efficacy of such resource allocation. Turtles with limited home ranges might be re-identified at a later date by physical searching or trapping (i.e., if they are carrying a passive integrated transponder tag (PIT tag) or shell markings) or by radio-telemetry. For wider ranging species (eg, highly migratory aquatic species), monitoring may be accomplished with satellite telemetry, which is very expensive, and its duration is limited to the battery life of the tag, often one to two years. Shell marking and PIT tags also require opportunistic happenstance (needle in a haystack), where a human is present at the right place and right time to detect the turtle (eg, on a nesting beach with a PIT tag scanner). A more recently utilized technology for wide ranging aquatic species is acoustic tags, which emit a high frequency sound at times intervals, and which have battery lives of up to ten years depending on the model. Such tags are detected by an array of underwater receivers, much like an automobile toll transponder. Acoustic tags have conventionally been attached externally to turtles, which puts them at risk for pre-mature detachment.[14] The author's research group recently gained permission for a pilot study to surgically implant subcutaneous acoustic tags in loggerhead turtles. Four surgeries have been completed to date; turtles healed without incident, and were released after an eight week monitoring period. Detection of their tags by the greater Atlantic acoustic receiver array is expected, but will not be known until receiver data is downloaded later this year. The potential for years-long post-release monitoring of rehabilitated turtles is exciting.

AZA SAFE AMERICAN TURTLES PROGRAM

The Association of Zoos and Aquariums (AZA) has developed the Saving Animals from Extinction program (SAFE). One of these programs, the American Turtles Program, seeks to end the illegal trade in native turtle species, especially highly prized, endangered species. The program seeks to provide veterinary intervention after confiscation of turtles by wildlife officials, followed by genetic analysis to determine their geographic origin, with the hope of eventual repatriation. Such programs are complex, and are influenced by infectious disease considerations, and lack of geographic clarity, in some cases. Interested parties may contact the author for further information.

REFERENCES

1. Lecis R, Paglietti B, Rubino S, et al. Detection and characterization of Mycoplasma spp. and Salmonella spp. in free-living European tortoises (Testudo hermanni, Testudo graeca, and Testudo marginata). J Wildl Dis. 2011; 47(3): 717-724.

1. Innis C, Kennedy A, Wocial J, et al. Comparison of oxytetracycline pharmacokinetics after multiple subcutaneous injections in three sea turtle species. J Herp Med Surg. 2020;30(3): 142-147.

2. Hawkins SJ, Cox S, Yaw TJ, et al. Pharmacokinetics of subcutaneously administered hydromorphone in bearded dragons (*Pogona vitticeps*) and red-eared slider turtles (*Trachemys scripta elegans*). Vet Anaes Analges. 2019;46(3): 352-359.

3. Royal LW, Lascelles BD, Lewbart GA, et al. Evaluation of cyclooxygenase protein expression in traumatized versus normal tissues from eastern box turtles (*Terrapene carolina carolina*). J Zoo Wildl Med;43(2): 289-295.

4. Norton TM, Clauss T, Sommer R, et al. Pharmacokinetic behavior of meloxicam in loggerhead (*Caretta caretta*), Kemp's ridley (*Lepidochelys kempii*), and green (*Chelonia mydas*) sea turtles after subcutaneous administration. J Zoo Wildl Med. 2021;52(1): 295-299.

5. Harms CA, Ruterbories LK, Stacy NI, et al. Safety of multiple-dose intramuscular ketoprofen treatment in loggerhead turtles (*Caretta caretta*). J Zoo Wildl Med. 2021;52(1): 126-132.

6. Cerreta AJ, Masterson CA, Lewbart GA, et al. Pharmacokinetics of ketorolac in wild eastern box turtles (*Terrapene carolina carolina*) after single intramuscular administration. J Vet Pharm Ther. 2019;42(2): 154-159.

7. Gregory TM, Harms CA, Gorges MA, et al. Pharmacokinetics of ketorolac in juvenile loggerhead sea turtles (*Caretta caretta*) after a single intramuscular injection. J Vet Pharm Ther. 2021;44(4): 583-589.

8. Raweewan N, Chomcheun T, Laovechprasit W, et al. Pharmacokinetics of tolfenamic acid in green sea turtles (*Chelonia mydas*) after intravenous and intramuscular administration. J Vet Pharm Ther. 2020;43(6): 527-32.

9. Corum O, Atik O, Corum DD, et al. Pharmacokinetics of tolfenamic acid in red-eared slider turtles (*Trachemys scripta elegans*). Vet Anaesth Analges. 2019;46(5): 699-706.

10. Feldman M. Prostaglandin induction technique. Available at: https://www.inducingturtles.com. Accessed August 13, 2021.

11. García-Párraga D, Crespo-Picazo JL, de Quirós YB, et al. Decompression sickness ('the bends') in sea turtles. Dis Aq Org. 2014;111(3):191-205.

12. Portugues C, Crespo-Picazo JL, García-Párraga D, et al. Impact of gas emboli and hyperbaric treatment on respiratory function of loggerhead sea turtles (*Caretta caretta*). Cons Physiol. 2018;6(1): cox074.

13. Hunt KE, Merigo C, Burgess EA, et al. Effects of ground transport in Kemp's ridley (*Lepidochelys kempii*) and loggerhead (*Caretta caretta*) turtles. Integr Org Biol. 2020;2(1):obaa012.

14. Smith BJ, Selby TH, Cherkiss MS, et al. Acoustic tag retention rate varies between juvenile green and hawksbill sea turtles. Anim Biotel. 2019;7(1): 1-8.

Ileocolic Intussusception in a Red-tailed Boa Associated with a Multicentric Round Cell Tumor

Tom Hellebuyck, Dr. med. vet., Dipl ECZM (herp), Jules Simard, Dr. med. vet., Richard Ducatelle, Prof. Dr. med. vet.

Affiliation: From the Department of Pathology, Bacteriology and Avian Diseases, Division of Poultry, Exotic Companion Animals, Wildlife and Experimental Animals, Salisburylaan 133, 9820 Merelbeke, Belgium

Session #2023

Key words: boa constrictor, ileocolic, intussusception, neoplasia, round cell tumor

Abstract: A 5-year-old, female red-tailed boa (*Boa constrictor constrictor*) was presented with weight loss, hyporexia and focal distention of the posterior coelom for two months. Physical examination revealed a firm, well-demarcated and movable intracoelomic mass in the posterior coelom but no other abnormalities. Ultrasonography revealed intestinal intussusception and a heterogeneous nodular aspect of the liver parenchyma. A complete blood cell count and serum biochemistry blood test results demonstrated lymphocytosis, hyperuricemia, hyperphosphatemia with an inverted Ca/P balance.[1] Explorative coeliotomy was performed and revealed anterograde invagination of the small intestine in the colonic lumen through the ileocolic junction with a length of approximately 14 cm. The intussusception was surgically repositioned using gentle digital manipulation. One month post-operatively, the snake died suddenly despite showing a good general condition and fecal production following assisted feeding. During necropsy, hepatomegaly, splenomegaly and renomegaly with the presence of multifocal, white nodules throughout the parenchyma were observed. Histopathological examination demonstrated infiltration of a malignant round cell tumor in the liver, spleen, kidneys and intestinal wall. Immunohistochemical stains (CD3, CD20, MAC387, S100 and NSE) were performed but could not confirm the cell origin of the multicentric round cell tumor. The present case highlights the need to consider round cell tumors as a primary cause of ileocolic intussusception in red-tailed boas. Cytological examination or histopathologic examination of organ biopsies might have provided a pre-operative diagnosis of a neoplastic process for the present case.

REFERENCES

1. Gibbons PM, Whitaker BR, et al. Chapter 35: Hematology and biochemistry tables. In: Divers SJ, Stahl SJ (editors). Mader's reptile and amphibian medicine and surgery. 3th ed., Elsevier, Saint Louis, 2019; 333.

Evaluation of dexmedetomidine-ketamine-midazolam combination in Ornate box turtles (*Terrapene ornata ornata*)

David Eshar, DVM, Dipl ABVP (ECM), Dipl ECZM (SM & ZHM), Theresa A. Rooney, DVM, MPH, Sara Gardhouse, DVM, Dipl. ACZM, ABVP (ECM) and Hugues Beaufrere, DVM, PhD, Dipl. ACZM, ABVP (Avian), ECZM (Avian)

Affiliation: From the Department of Clinical Sciences, College of Veterinary Medicine, Manhattan, KS 66502 USA (Eshar, Rooney, Gardhouse) and the Department of Clinical Studies, Ontario Veterinary College, University of Guelph, Guelph, ON N1G *Canada (Beaufrere)*

Abstract:
16 ornate box turtles, were injected with dexmedetomidine (0.1 mg/kg), midazolam (1 mg/kg) and ketamine (10 mg/kg) administered intramuscularly in the front limbs. Time to first response, time to maximal effect, the plateau phase, and time to recovery from reversal administration were recorded. Vitals, muscle tone in all limbs and the neck and reflexes (palpebral, withdrawal, and toe pinch) and endotracheal intubation were recorded every 5 minutes and until fully recovered. Response to an intramuscular injection of 0.1 ml sterile saline was assessed every 15 minutes. Atipamezole (0.5mg/kg, IM) and flumazenil (0.05mg/kg, SC) were administered in the front limb 60 min after the initial DKM injections. The mean time to first response, time to maximal effect, the plateau phase, and time to recovery were 2.1 minutes, 14.9 minutes, 38.7 minutes, 7.8 minutes, respectively. Respiratory rate was not observed during the study in any animals. The body temperature significantly increased over time. The palpebral reflex was persistent in 43% of throughout the immobilization event. The tail pinch reflex remained intact for 13% of animals throughout the immobilization event when the toe pinch reflexes were apparently absent. All turtles recovered from their immobilization event with no adverse effects noted during or after their anesthesia. This study showed that this DKM protocol may be safely used in ornate box turtles to produce rapid-onset, deep sedation or light anesthesia for roughly 40 minutes, long enough to perform many clinical procedures or to facilitate intubation for inhalant anesthesia, followed by a smooth, reliable recovery.

Keywords: Anesthesia, chelonian, dexmedetomidine, ketamine, midazolam, *Terrapene ornata ornata,* ornate box turtle

Scales and Sarcomas: The Diagnosis and Treatment of Neoplasia in the Reptile Patient

Jane Christman, DVM[1]

[1]Detroit Zoological Society, Royal Oak, Michigan, United States of America

Introduction

Reptile patients are subject to neoplasia at a similar frequency and prevalence as other taxa, including birds and mammals.[1,2] However, very little is known about the relative frequency of encountering certain neoplasms in reptilian patients. Reptile owners are becoming more and more interested in advanced diagnostic and treatment recommendations for their pets, and thus veterinarians treating reptiles should be familiar with current literature and recommendations for approaching neoplasia in these unique patients. Largely, the clinical approach to a reptile patient with suspected neoplasia follows a similar process as mammalian patients and working with a board-certified veterinary oncologist whenever possible can allow clinicians to pursue additional medical and surgical options to provide the best clinical outcome.

The prevalence of neoplasia in reptiles varies amongst species. A recent literature review of tumors in reptiles from 2007-2017 found that snakes have the highest incidence of neoplasia, followed by lizards. Chelonians had the lowest incidence of neoplasia with fewer published cases identified in the literature.[1] The reason for the variation is unknown but could be due to relative lifespan of the species or over representation in species kept commonly as pets. All body systems can be affected by neoplastic disease. Across literature representing reptilian neoplastic disease across species, the hematopoietic system is the most common body system affected, followed by hepatobiliary and integumentary systems.[1]

The majority of neoplastic disease in reptiles lacks an underlying etiology, however there are a few exceptions. Fibropapilloma is a common tumor of marine turtles, is associated with chelonid herpesvirus 5.[3] In addition, the identification of C-type virus particles within a chondrosarcoma from a corn snake (*Elpahe guttata*), suggests a possible underlying viral etiology, as well as retrovirus isolated from boid snakes with inclusion body disease.[1,4,5] Genetic and hereditary tumors are considered uncommon in reptiles. However, a chromatophoroma (iridophoroma) of leopard geckos (*Eublepahrus macularis*) of the lemon frost color morph with suspected heritability was recently described.[6] Additional examples include renal adenocarcinoma, which occurred in three genetically related cape coral snakes.[7]

Diagnosis

In all oncology, early diagnosis of neoplastic disease is more likely to lead to a better clinical outcome. The diagnosis of neoplasia in a reptilian patient involves thorough evaluation of the patient's history and clinical signs. Oftentimes, reptile patients present with a visible mass or lesion identified on palpation and physical exam. Differential diagnoses for masses on reptiles include inflammatory disease (abscess, granuloma), tissue hypertrophy, hypovitaminosis A, squamous metaplasia, traumatic and degenerative processes in addition to neoplasia, and a thorough work-up is recommended to rule other etiologies first.[2]

Routine hematology and biochemistry are considered baseline diagnostics for staging reptiles with neoplasia. Values should be interpreted in comparison to published reference ranges for the presenting species. Oftentimes, bloodwork is within normal limits and no obvious abnormalities are detected. Debilitated reptiles can show nonspecific bloodwork changes, including hypercalcemia, hypoglycemia, hyperphosphatemia and increased plasma CK.[1,2]

The use of diagnostic imaging, including radiographs and ultrasound, are especially helpful for the characterization of a suspected neoplasm in a reptile patient. Three-dimensional imaging such as computerized tomography (CT) or magnetic resonance imaging (MRI) are becoming more and more common and can help eliminate superimposed structures and enhance detail for determining the extent of disease and evaluating for local and distant site metastasis.[2]

Treatment

A comprehensive treatment plan for a reptile presenting with neoplasia includes either surgical or medical therapeutic intervention. An individualistic approach, combined with the treatment goals of the caretaker, should be taken into consideration. Quality of life should be the ultimate priority when coming up with a treatment plan. Ensuring proper husbandry throughout the treatment period is essential, including proper hydration, diet, humidity, and maintenance within preferred optimum temperature zones. Supportive care measures, including application of analgesia, and supplemental or support feeding should be considered, depending on the clinical state of the patient, and treatment modalities considered.[1,2]

Surgery is the most commonly used therapeutic modality for the treatment for tumors in reptiles and used alone or in combination with chemotherapy can increase efficacy of treatment and has the potential to lead to clinical resolution. Evaluation of the patient and tumor type for suitability of a surgical approach is necessary, including size and location. Excisional techniques depend on treatment goals and tumor location and can include debulking, marginal incision or wide excision.[2] Excisional surgery should be followed with histopathology and review by a pathologist with knowledge of reptiles.

Medical interventions can include the use of chemotherapy and radiation therapy. Chemotherapy consists of a chemical agent that targets rapidly proliferating tissue and can be applied locally or systemically.[8] As reptiles have a lower basal metabolic rate than mammals and birds, there is potential to be more sensitive to chemotherapy at standard doses based on surface area.[1] Almost no prospective research has been performed on the use of chemotherapeutic agents in reptiles. Consultation with a veterinary oncologist is recommended for dosing chemotherapeutic agents in reptiles.

Types of chemotherapy include local or systemic applications. Intratumoral chemotherapy for exposed tumors of the skin is commonly employed due to ease of administration at the tumor site but should not be used for large or bulky tumors.[1] Systemic chemotherapy can be challenging due to difficulties gaining intravenous access in reptiles, especially if the patient is debilitated or requires frequent IV access. The use of vascular access ports (VAP) can be considered and are commercially available for larger reptiles that require frequent venous access.[1,2] Monitoring hematology parameters throughout treatment with chemotherapy is necessary, as the timeline for development of leukopenia secondary to bone marrow suppression with chemotherapeutic protocols has not yet been established.

Radiation therapy can be delivered at a variety of different protocols for palliative or curative intent. Reports of the use of radiation therapy are sparse in the literature. Side effects can be acute or delayed include vasculitis, cellulitis, dysecdysis, site-specific toxicity and death.[9] The toxic dose of radiation is species dependent, with snakes having an LD50 dose similar to that of humans, and lower than that of other reptile taxa.[9] The small size of most pet reptiles makes a focused beam of radiation challenging and may increase the incidence of systemic side effects.[1] Speculation as to the tolerance of reptiles based on physiology to radiation therapy and the efficacy in treatment of different tumor types in comparison to mammals and birds is unknown.

References

1. Christman JE, Devau M, Wilson-Robles H, Hoppes S, Rech R, Russel KE, Heatley JJ. Oncology of Reptiles: Diseases Diagnosis and Treatment. 2017. *Veterinary Clinics of North America: Exotic Animal Practice*. 20(1): 87-110.
2. Mayer J, Moore AS. Chapter 78: Oncology. In: *Mader's Reptile and Amphibian Medicine and Surgery. 3rd ed.* 2019 Elsevier. 827-834.
3. Lackovich JK, Brown DR, Homer BL, Garber RL, Mader DR, Moretti RH, Patterson AD, Herbst LH, Oros J, Jacobson ER, Curry SS, Klein PA. Association of herpesvirus with fibropapillomatosis of the green turtle *Chelonia mydas* and the loggerhead turtle *Caretta caretta* in Florida. *Diseases of Aquatic Organisms.* (1999) 37:89–97.
4. Lunger PD, Hardy Jr WD, Clark F. C-type virus particles in a reptilian tumor. *Journal of the National Cancer Institute.* 1974. 52(4):1231-5.
5. Jacobson ER, Orós J, Tucker SJ, Pollock DP, Kelley KL, Munn RJ, Lock BA, Mergia A, Yamamoto JK. Partial characterization of retroviruses from boid snakes with inclusion body disease. *American journal of veterinary research*. 2001 62(2):217-24.
6. Szydłowski P, Madej JP, Duda M, Madej JA, Sikorska-Kopyłowicz A, Chełmońska-Soyta A, Ilnicka L, Duda P. iridophoroma associated with the Lemon frost colour morph of the leopard gecko (Eublepharis macularius). *Scientific Reports.* 2020;10(1):1-6.
7. Keck M, Zimmerman DM, Ramsay EC, Douglas M, Reavill DC. Renal adenocarcinomas in cape coral snakes (Aspidelaps lubricus lubricus. *Journal of herpetological medicine and surgery.* 2011; 21(1):5-9.
8. Moore AS, Mayer J, Harrison TM. Chapter 123: Cancer Chemotherapy. In: *Mader's Reptile and Amphibian Medicine and Surgery*. 3rd ed. 2019. Elsevier. 1177-1181.
9. Nagata K, Mayer J. Chapter 124: Radiation Therapy. In: *Mader's Reptile and Amphibian Medicine and Surgery*. 3rd ed. 2019. Elsevier. 1182-1184.

Herpesvirus Associated Cheilitis in Two Panther Chameleons (*Furcifer pardalis*)

Tom Hellebuyck, Dr. med. vet., Dipl ECZM (herp), Liesbeth Couck, Wim Van den Broeck, Prof. Dr. med. vet., Richard Ducatelle, Prof. Dr. med. vet., Rachel E. Marschang, Dr. med. vet., Dipl ECZM (herp)

Affiliation: From the Department of Pathology, Bacteriology and Avian diseases, Faculty of Veterinary Medicine, Ghent University, Salisburylaan 133, B-9820 Merelbeke, Belgium (Hellebuyck, Simard, Ducatelle), the Department of Morphology, Faculty of Veterinary Medicine, Ghent University, Salisburylaan 133, B-9820 Merelbeke, Belgium (Couck, Van den Broeck), LABOKLIN GmbH & Co KG, Steubenstrasse 4, DE-97688 Bad Kissingen, Germany (Marschang)

Session #2025

Abstract: Intracellular epidermal inclusions were detected within histological sections of skin biopsies obtained from two panther chameleons (*Furcifer pardalis*) with chronic cheilitis. Transmission electron microscopy (TEM) confirmed the abundant presence of icosahedral intracytoplasmic and intranuclear viral particles in infected keratinocytes with an average diameter of 120-125 nm consistent with herpesviruses (HVs). In addition, TEM revealed intercellular migration and migration of cytoplasmic virions into keratinocyte nuclei and features suggestive of capsid assembly, nuclear egress with primary envelopment as well as anterograde transport leading to virion assembly. Polymerase chain reaction primers targeting a conserved region of herpesvirus DNA-dependent DNA polymerase were used to amplify and sequence a product from skin biopsies of both chameleons.[1] Analysis of available sequence information showed that the detected HV in both chameleons originated from the lineage of Alphaherpesvirinae, which is referred to as chamaeleonid herpesvirus 1 (chamHV 1). Additional sequence data is necessary to clarify the relationships between the HV found in the chameleons and other HVs and to understand the evolution of this virus. Based on the identical findings in both chameleons, chamHV 1 is considered to be an etiological candidate for the development of cheilitis in panther chameleons. This is the first report of skin lesions in a chameleon species associated with HV infection.

Key words: chameleon, cheilitis, *Furcifer pardalis*, herpesvirus, transmission electron microscopy

REFERENCES

1. Van Devanter DR, Warrener P, et al. Detection and analysis of diverse herpesviral species by consensus primer PCR. J Clinic Microbiol. 1996; 34(7): 1666-1671.

Reptile Cardiology

Lionel Schilliger, DVM, DECZM (Herpetology), DABVP (Reptile & Amphibian Practice)

Affiliation : From the Clinique Vétérinaire du Village d'Auteuil, 35 rue Leconte de Lisle, 75016 Paris, France.

Session #ExoticsCon Virtual Conference 2021 (ARAV)

Abstract**:** Reptilian hearts can be loosely classified as being crocodilian or non-crocodilian hearts. In non-crocodilian reptiles, the ventricle is not physically divided but pulmonary and systemic blood flow are separated and regulated by functional plasticity. In these species, oxygenated and deoxygenated blood flowing from the aortic arches is partially mixed, the amount of mixing being determined by the degree of muscular ridge and interventricular septum development. The contraction of the muscular ridge during systole and of the interventricular septum during diastole serves to form an almost complete separation of pulmonary and systemic blood flow, particularly in monitor lizards and pythons. In cases of anoxia (e.g. prolonged diving, swallowing large preys, respiratory distress or hibernation), vasoconstriction of the pulmonary arteries/artery causes a high pulmonary outflow resistance. Together with a decreased pressure in the pulmonary veins, a right to left intracardiac shunt occurs and blood from the right atrium shifts towards both aortic arches, instead of the pulmonary trunk. This ensures bypass of the lungs and blood perfusion to vital organs during anoxia. The cardiac anatomy and physiology in crocodilian species resembles that of mammals. Indeed, the ventricles are physically divided by an intraventricular muscular septum with divided pulmonary and systemic circulation. However, during apnea, the crocodilian heart is able to function in a manner similar to that of non-crocodilian reptiles with a right to left shunt. This unique versatility is made possible by the foramen of Panizza, an interventricular septal opening that connects the left and right aortic trunks, which, coupled with an active control of pressure gradients in the ventricles, allows for mixing of oxygenated and deoxygenated blood. Whilst the degree of shunting is known for many species, the functional mechanisms and physiologic role of cardiac shunting have been discussed extensively with no final or conclusive interpretations.

Until recently, primary cardiovascular conditions were rarely considered to affect ectotherms. This assumption is now challenged by the increased availability of diagnostic tools that have improved our knowledge of reptile cardiovascular anatomy, physiology and pathology, such as auscultation, electrocardiography, radiography, and in particular, ultrasound. Although the pharmacokinetics and pharmacodynamics of cardiovascular drugs are poorly understood in these animals, basic principles remain applicable; these include pharmacologic (e.g. diuretics, vasodilators, inotropes, angiotensin-converting enzyme inhibitors) and non-pharmacologic interventions (e.g. reducing stress, maintaining the reptile at the lower end of its preferred optimum temperature zone, avoiding overfeeding and stress). Further research is needed to establish species-specific cardiac reference ranges and evidence-based treatment options.

Key words: Reptile - Cardiology - Ultrasound - Electrocardiography – Cardiac drugs.

INTRODUCTION

Despite a good basic understanding of reptilian cardiac anatomy and physiology, cardiology remains a poorly developed sub-specialty in herpetological medicine when compared to other disciplines. Translating our present knowledge of physiology and pathology into clinical practice will allow us to advance the discipline.

TOPOGRAPHICAL ANATOMY

In snakes, the heart is found along the midline axis and its longitudinal position varies with species. In non-tree dwelling snakes, it is found at about 25% of total body length, and about 15% (i.e. more cranially) in arboreal species. In marine and freshwater snakes, it is positioned near the middle of the body (25 to 45% of total body length).[1,2] A cranially positioned heart reduces the hydrostatic pressure above the heart and so stabilizes variations in cephalic blood pressure.[3] In the coelom, the snake heart is found adjacent to the caudal tracheal rings, caudal to the thyroid, cranial to the bronchial bifurcation, close to the cranial pole of the lung(s) and slightly cranial to the liver.[4] Externally, the position of the heart is indicated by percutaneous visualization of the ventral precordial tap by placing the animal in dorsal recumbency and stretching it. This is particularly obvious in the anesthetized and relaxed reptile.

The chelonian heart is located in the ventral midline at the level of the intersection of the humeral, pectoral and abdominal scutes of the plastron. In some species it is positioned more caudally, between the junction of the humeral-pectoral and pectoral-abdominal scute lines. This may however be an unreliable landmark as the position of the pectoral and humeral scutes differ widely amongst species. In the coelom, the heart is bordered dorsally by the lungs, laterally by the liver lobes and ventrally by the acromion and coracoid processes.[5] In soft-shelled turtles of the *Tryonychidae* family, which possess a very streamlined carapace and hence small coelomic cavity, the heart is displaced to the right. This is due to long protractile neck, which when retracted into the carapace causes a displacement of the heart and liver to the right and of the stomach to the left.[2]

In most species of lizards, the heart is encased in the pectoral girdle, in the gular region. However, in Tegus (*Teiidae*), Gila monsters (*Helodermatidae*) and monitor lizards (*Varanidae*), the heart is located more caudally in the coelomic cavity, behind the cranial part of the sternum.[6]

In crocodilians, the heart lies in the body midline. The pericardial sac occupies the caudal ventral part of the thoracic cavity. The apex lies deep between the right and left medial aspect of the hepatic lobes and apposes the liver isthmus. The base of the heart extends slightly cranial to the cranial border of the liver and lies within the mediastinum, ventral to the primary bronchi and esophagus.[7]

EXTERNAL GROSS MORPHOLOGY

Despite minor variations, the external anatomy of the reptilian heart tends not to vary significantly between species. The non-crocodilian heart has historically been described as a three-chambered organ. Because the sinus venosus (see below) is now considered as an additional chamber, the non-crocodilian heart can be considered as an atypical four-chambered heart.

As in all vertebrates, the reptilian heart is completely covered by a white fibrous pericardial sac and lies within a cœlomic space, the pericardial cavity. The heart apex is attached to the pericardium by a ligament termed the *gubernaculum cordis*, except in snakes and varanids in which the ventricle remains free within the pericardial space. Within the pericardial sac, the heart is covered by the epicardium, which forms the serous lining of the pericardial cavity.[1]

Reptilian hearts can be classified into two basic patterns: the **non-crocodilian heart** (in squamates and chelonians) and the **crocodilian heart** (in crocodiles, alligators, gavials and caimans) (Figure 1).

Non-crocodilians

Seen from the outside, the heart of non crocodilian reptiles is composed of:

- **Four main cavities**: a single ventricle (similar to an inverted cone, it is broad and flattened dorsoventrally in many Testudines or elongated and oval in snakes, with rounded edges), two atrial chambers which can be seen on a ventral view (the right atrium being, especially in snakes and in Testudines, more developed than the left one, whereas they are of equal size in varanids) and a sinus venosus, in which cardiac contraction is initiated. This sac-like structure of tubular or triangular shape is only visible from a dorsal view of the heart, and rests on the dorsal aspect of the right atrium, at the confluence of the three caval veins in most reptiles: the right precaval vein, the left precaval vein and the postcaval vein. The ventral wall of the sinus venosus is attached to the dorsal wall of the ventricle by the dorsal ligament. All the walls of the sinus venosus are thinner than those of the other cardiac cavities. The atria are separated from the ventricle by quite a deep oblique-angled sulcus, called the coronary groove (*Coronary Sulcus*).[1]

- **Three arterial trunks**, running from the ventricle between the two atria, rotating towards the right, and forming an angle of 180°, typical of reptiles: the left aortic arch, right aortic arch and pulmonary trunk. The two aortic arches merge together caudally to form the common abdominal aorta (dorsal aorta). The right aortic arch coming from the left side of the ventricle develops into either one or two common carotid arteries. The pulmonary trunk divides up into the left and right pulmonary arteries in reptiles with two functional lungs (e.g. boid snakes). In others, the pulmonary trunk develops into one anterior pulmonary artery pointing forwards and a posterior pulmonary artery pointing backwards.[1]
- **Four afferent veins**: the pulmonary vein, the two precaval veins and the post-caval vein, draining into the sinus venosus.[1]
- **A well-developed coronary vascularization** that arises from the systemic carotid trunk and supplies the compact myocardium, the sinus venosus and the two atria.[1]

Crocodilians

The heart of crocodilians has **two atria** and **two ventricles, physically separated** by a distinct muscular septum into right (pulmonary + systemic) and left (strictly systemic) sides. On gross examination, the heart is comparable that of mammals and birds, except for the presence of the left and right aortic arches. The *gubernaculum cordis* extends from the right ventricle (not from the apex as in Testudinae and Sauria) and anchors the heart to the pericardium. The right and left atria are of similar size. The sinus venosus resembles an elongated pyramid, with the base lying cranially and apex caudally. In the saltwater crocodile (*Crocodylus porosus*), the sinus venosus is formed is formed by the union of the hepatic vein, postcaval vein and right precaval vein.[1]

Three major vessels exit the ventricle: the **right aorta** leaves the left ventricle and gives off cranially the carotid and subclavian arteries. It continues caudally as the dorsal aorta. The **pulmonary artery** leaves the right ventricle and supplies the left and right lungs. The **left aorta** also leaves the right ventricle. As they exit the ventricle, all three vessels are of similar diameter. As they run cranially, the aortae and the pulmonary artery are bound together by a connective tissue sheath.[1]

INNER STRUCTURES

Non-crocodilians

Seen from the inside, the heart is composed of:

- Three ventricular sub-chambers, recessed within the ventricular cavity and communicating with each other: the **cavum venosum** (the right dorsal chamber of the ventricle), the **cavum arteriosum** (the left dorsal chamber of the ventricle) and the **cavum pulmonale** (left chamber or "ventral ventricle"). Reptiles belonging to the genus *Varanus* possess an enlarged cavum arteriosum and reduced cavum venosum. This gives the heart a symmetrical appearance, in contrast to other squamates. Thus, two kinds of cardiac morphologies can be roughly defined in non-crocodilian reptiles:

• **The typical non-crocodilian heart** has a ventricle which is a continuous single chamber, operating as a single pump during the entire cardiac cycle. This is seen in most chelonians such as freshwater turtles and tortoises, and in squamates including aquatic snakes, grass snakes, rattlesnakes, and most lizard species. In this type of heart, the muscular ridge is relatively small and not well developed. The right aortic and the intraventricular pressures are equal.[1]

• **The varanid and python heart** possesses a large, well developed muscular ridge, which almost forms a complete ventricular septum that can be visualized as an asymmetric thick wall. This is also found in at least one species of marine turtle, the leatherback turtle. Moreover, in these reptiles, the *cavum arteriosum* is enlarged and the *cavum venosum* is reduced. In this heart model, the *cavum pulmonale* becomes functionally separated from the *cavum venosum* and from the *cavum arteriosum* during systole, Thus, this kind of ventricle functions as a double pump, with higher pressures in the *cavum venosum* and the systemic arches than in the *cavum pulmonale* and pulmonary arteries. This complete separation of the pulmonary and the systemic circulations has been clearly shown in varanids (*Varanus exanthematicus, Varanus niloticus*), in the Burmese python (*Python molurus bivittatus*) and in the ball python (*Python regius*).[1,8,9,10]

- **Several openings and valves**: the pulmonary vein drains into the dorsal wall of the left atrium through the valve-free pulmonary vein orifice, located close to the interatrial septum. The sinus venosus proceeds into the right atrium at the sinoatrial opening. This orifice has a pair of flap valves derived from the endocardium that are membranous or muscular, depending upon the species, and are known as sinoatrial valves. The atrial chambers communicate with the single ventricle through the left and right atrioventricular orifices. These orifices are both equipped with a single atrioventricular valve, known as the single septal monocuspid atrioventricular valve. The left and right aortic arches arise from the cavum venosum at two orifices with bicuspid valves, rather than tricuspid valves as in mammals. The pulmonary trunk is a continuation of the cavum pulmonale and its base also contains small semilunar valves.[1]
- **two main septa**: one vertical septum, so called "interventricular septum", located in the ventricular cavity, between the cavum arteriosum and the cavum venosum. Because of its position facing the interatrial septum, it contributes to the separation of pulmonary veinous blood from systemic flow during ventricular diastole. A second relatively strong muscular ridge, the muscular ridge (also called horizontal septum), originates between the pulmonary trunk and the left aortic arch, and twists towards the dorso-apical region of the ventricle. This ridge marks a separation between the cavum venosum and the cavum pulmonale. It is more developed in varanids and pythons compared to other squamates and most chelonians (with the exception of sea turtles and giant tortoises).[1]
- **another septum**, the bulbuslamella**,** which lies opposite to the muscular ridge, curving over and around it. It is situated immediately to the right of the vertical septum.[1]

Crocodilians

Seen from the inside, the crocodilian heart is composed of:

- a **left and right ventricle**, separated by an interventricular septum.
- **several openings and valves**: the base of each efferent arteries has a bicuspid valve, the medial cusp being longer than the lateral one. Significantly, the medial cusp of the right aorta covers the foramen of Panizza during ventricular systole. The bicuspid atrioventricular valves, which guard the connection between the atria and ventricles, are composed of a left septal and a right marginal cusp (they are bicuspid valves). As in mammals, the pulmonary veins are devoid of valves. The right and left aorta are connected via the foramen of Panizza and the dorsal connecting artery. The former structure lies just above the bicuspid valves at the base of both aortas. The dorsal connecting artery is located more distally but is rostral to the convergence of both aorta into the single aorta.[1,7]

PHYSIOLOGY

Conduction and heart rate

Unlike mammals, reptiles do not have a specialized cardiac conduction system such as pacemaker nodes and Purkinje fibers. Instead, the contractions are initiated by cardiac muscle. The contractions originate in fibers within the *sinus venosus* of the right atrium, which spread in sequential coordination, first to the left and then towards the apex caudally. The ventricle is depolarized starting at its base and then proceeding to the left. Repolarization starts from the base and spreads symmetrically to the right and left towards the apex. It is believed that specific areas of cardiomyocytes slow conduction between different heart compartments, creating marked delays in electrical conduction. Such delays are clearly visible on the ECG of all researched reptile species with a characteristic rapid systolic phase and far slower diastolic one. Thus, the spongy myocardium of the ventricle functions as a medium for both conduction and contraction.[1,2]

The heart receives innervations from both parasympathetic and sympathetic fibers. The parasympathetic fibers run in the vagus nerve and provide cholinergic (inhibitory) control. The less developed sympathetic fibers cause positive chronotropism and inotropism via adrenergic innervation.[11,12]

The heart rate (HR) is generally slower in reptiles than in mammals or birds. HR is dependent on numerous factors. Myocardial efficiency is optimum when the reptile is within its preferred optimal temperature zone and at its preferred body temperature. Tachycardia occurs when the animal is exposed to higher temperatures, and, together with peripheral vasodilatation, increases heat loss from cutaneous vascularization. Conversely, lower temperatures stimulate bradycardia, and, together with peripheral vasoconstriction, reduce heat loss. This mechanism, which often anticipates internal temperature variations, is regulated by cutaneous thermoreceptors. HR is proportional to metabolic level and inversely proportional to body size. Reptiles that experience hypovolemia (e.g. as a result of surgery or trauma) may become tachycardic to maintain cardiac output and tissues oxygenation. Bradycardia is observed during apnoea as pulmonary resistance increases and blood flow to the lungs decreases effectively creating a right to left cardiac shunt.37 Other effects on HR include digestion, gravidity, sensory stimulations (such as handling, postural and gravitational stress), méthylatropine and certain anesthetics and sedatives (e.g. alpha-2 agonists).[2]

HR can be estimated using an allometric formula; HR (beats per minute) = $33.4xW(kg)^{-0.25}$ for non-ophidian species and HR (beats per minute) = $20.6xW(kg^{)-0.229}$ for ophidians.[13] This formula should be used in conjunction with known HR parameters (if available), as it tends to over-estimate HR in some species such as bearded dragons.[14]

Intracardiac blood flow during normal breathing

Non-crocodilians (Figure 2)

Oxygenated and deoxygenated blood flowing from the aortic arches are partially mixed, the amount of mixing being determined by the degree of development of the muscular ridge and interventricular septum.

During atrial systole, both atrioventricular valves open simultaneously, blocking the interventricular canal between the cavum arteriosum and the cavum venosum. The deoxygenated blood contained in the right atrium first fills the cavum venosum and the cavum pulmonale, incompletely separated by the muscular ridge, and the oxygenated blood contained in the left atrium then flows into the cavum arteriosum. Blood from the right atrium enters the cavum venosum slightly before blood from the left atrium fills the cavum arteriosum. Thus, during ventricular diastole, systemic and pulmonary blood flow are partially separated by the interventricular septum.[1,2]

During ventricular systole, both atrioventricular valves close the atrioventricular funnels and open the interventricular canal, allowing blood from the cavum arteriosum to fill the cavum venosum (once the cavum pulmonale is filled). The muscular ridge moves cranially and separates the cavum venosum and the cavum pulmonale. Oxygenated blood is then pumped out of the cavum venosum on to the left and right aortic arches and deoxygenated blood is pumped out from the cavum pulmonale towards the pulmonary arterial trunk. Because the diastolic blood pressure is lower in the pulmonary arches than in the systemic ones, blood flows first in the pulmonary arteries and then in the systemic arteries.[1,2] As already mentioned, the muscular ridge nearly completely separates the two sides of the ventricle in varanids and pythons. This produces a more pronounced separation of blood flow and allows for higher systemic pressures.[1,2,8,9,10]

Crocodilians

When the crocodilian has free access to air, the heart functions in a similar way to mammals with separate systemic and pulmonary circulations. Because the blood pressure in pulmonary vasculature is much lower than in the systemic vasculature, the pressure generated in the right ventricle is much lower than in the left ventricle. The blood pressures in the left and right aorta are equalized via the foramen of Panizza (which opens during ventricular diastole) and the dorsal connecting artery. Consequently, the bicuspid valve at the orifice of the left aorta (remember that this exits the right ventricle) remains closed, as the right ventricular pressure does not exceed left aortic blood pressure. Hence, all the blood in the right ventricle is pumped into the pulmonary circulation. In addition, the blood in the left ventricle goes primarly into the right aorta as one of the valves guarding the right aorta folds back to cover the entire foramen of Panizza during systole.[1,2,7,15]

Intracardiac blood flow during anoxia (cardiac shuntings)

Non-crocodilians (Figure 3)

Non-crocodilian reptiles exhibit an interesting distinctive cardio-circulatory feature in that intracardiac blood flow may vary according to systemic blood oxygenation levels (PO2 levels) and to the relative vascular resistance of the pulmonary and systemic circuits. In cases of anoxia (prolonged underwater diving, slowly swallowing large preys, hibernation or respiratory distress), an increase in pressure of the pulmonary artery (due to vasoconstriction) and a decreased blood pressure in the pulmonary veins are observed, creating a high pulmonary outflow resistance. Subsequently, a right to left intracardiac shunt may occur and blood from the right atrium shifts towards both aortic arches, instead of the pulmonary trunk. This shunt ensures blood perfusion of vital organs during anoxia. While cardiovascular shunting is detrimental in birds and mammals and is only seen in congenital abnormalities, it serves important physiological functions in reptiles.[2,16,17]

This kind of cardiac shunting mechanism is observed in chelonians and also in most squamates, where the heart acts as a single chamber pressure pump. This is termed "pressure difference shunting", whereby the balance between R-L or L-R shunts depends essentially on the ventilatory state of the animal (i.e. the pulmonary and systemic vascular resistances). "Washout shunting" may also be responsible for cardiac shunting in reptiles. The main difference between these two shunting mechanisms depends on the ability of the muscular ridge to separate the *cavum pulmonale* from the *cavum venosum* and *cavum arteriosum* during systole. "Pressure difference shunting" only occurs in crocodiles, whereas in turtles, both "pressure difference shunting" and "washout shunting" occurs. The "washout" mechanism is probably the primary physiological determinant of intracardiac shunting in species with a partially developed "double pressure" pump (varanids and pythons). In varanids and pythons, the separation between the *cavum pulmonale* and the two other cava prevents any resistance in outflow pressure to influence intracardiac shunting during ventricular systole (as is the case in other squamates). "Washout shunting" occurs when the residual volume of blood remaining in the *cavum venosum* is mixed within the pulmonary or systemic circuit during the cardiac cycle. During right atrial systole, even though most of the systemic venous blood passes into the *cavum pulmonale*, a residual amount of deoxygenated blood remains in the *cavum venosum*. During ventricular systole, the *cavum arteriosum* injects blood into the *cavum venosum* and into the systemic arches. This results in the mixing of the two atrial supplies and thus creates a R-L shunt. At the end of ventricular systole, oxygenated pulmonary venous blood may remain in the *cavum venosum* (i.e. end-systolic volume of the *cavum venosum* is not zero). During the next diastole, de-oxygenated systemic venous blood pumped out from the right atrium through the *cavum venosum* and the *cavum pulmonale* will mix with this oxygenated pulmonary venous blood, resulting in a L-R intracardiac shunt. This end-systolic oxygenated blood is washed from the *cavum venosum* into the *cavum pulmonale* during the following diastole. Thus, the degree of R-L shunt depends on the volume of deoxygenated blood in the *cavum venosum* at end diastole in addition to the volume of blood ejected from the *cavum arteriosum* during systole. The degree of L-R shunt depends on the volume of oxygenated blood in the *cavum venosum* at the end of systole in addition to the volume of systemic venous blood. Thus, "washout shunting" is dependent on preload, contractility of the heart and afterload. This has been demonstrated in the ball python, in which shunting occurs in the *cavum venosum*, when the atrioventricular valves open but the muscular ridge is not yet closed. During fasting and resting, up to twice as much blood flows into the systemic circulation than into the pulmonary one (R to L shunt) whereas during digestion, more blood is pumped through the pulmonary circulation (L to R shunt) because of increased oxygen demand.[2,16,17]

The direction and degree of cardiac shunting in squamates and chelonians may also be influenced by internal and/or external factors, including temperature, disease/healing, digestion, vascular sphincters, hibernation and, in aquatic species, the dive reflex. Various investigative approaches have been followed in order to better understand the potential physiological benefits. It is also reasonable to assume that anesthetic drugs, pulmonary ventilation, and the composition of inhaled air may have an effect on cardiac shunting during anesthesia. This is likely to vary between species and may impact the duration of anesthetic recovery and anesthetic gas (if used) excretion.

Crocodilians

During apnea, the crocodilian heart functions similarly to a non-crocodilian heart with a right to left shunt, and up to 25% of blood flow may bypass the lungs. Dual mode of action is made possible by the foramen of Panizza, together with an active control of pulmonary vasculature outflow. When in apnea, an increase pressure of lung tissue together with cog-like valves in the pulmonary artery causes an increase in pulmonary resistance. These results in the pressure of the right ventricle overcoming that of the pressure in the left systemic aorta. Then, there can be simultaneous output from the right ventricle into the left aorta and the pulmonary artery. Most of the blood probably follows the course of the left aorta. However, some blood could pass into the right aorta via the foramen of Panizza and at the confluence of the left and right aortic arches (resulting in some mixing of oxygenated and deoxygenated blood in the systemic circulation).[2,15,16,17]

CARDIOVASCULAR DISEASES IN REPTILES

A variety of pathological lesions have been reported involving the reptilian cardiovascular system. Inadequate husbandry, resulting in chronic stress, immune suppression and malnutrition, is the most likely common predisposing factor for the development of cardiovascular conditions in captive reptiles. Cardiac diseases can be classified as non-infectious, infectious or parasitic.

Non Non infectious cardiac disease

Congenital defects

Congenital cardiac diseases are considered rare in reptiles. A secundum atrial septal defect located on the craniodorsal portion of the interatrial septum was reported in a Komodo dragon (*Varanus komodoensis*). Various congenital conditions have been reported in several species of snakes: ventricular mural hypoplasia with plasmacytic pericarditis (Burmese python, *Python molurus bivittatus*); aortic valvular stenosis with secondary cardiomyopathy (Children's python *Liasis childreni*); and enlarged/abnormally shaped cardiac chambers, bifid ventricles and reduced muscular ridge causing inadequate ventricular pressure separation with pronounced hemodynamic consequences (ball pythons, *Python regius*).[2]

Aortic and sub-aortic stenosis have been reported in other species: histopathological examination of the heart of a green iguana (*Iguana iguana*) with aortic stenosis associated with dilatation of the right atrium and the ventricle revealed atrophy of myocardial fibers and a thickening of the intima of both aortic arches with consequent narrowing of the lumen. No diagnosis was established but chronic congenital lesion was suggested as a differential diagnosis. Bilateral subaortic stenosis was also described in an alligator (*Alligator mississippiensis*).[2]

Cardiomyopathies

Cardiomyopathies have been described in various species of snakes. A black kingsnake (*Lampropeltis niger*) that presented with marked cardiomegaly, dilatation of the ventricle and congestive heart failure (CHF) was diagnosed with dilated cardiomyopathy upon necropsy.96 CHF was also reported in a Deckert's rat snake (*Pantherophis obsoleta deckertii*) with degeneration and necrosis of myocardial fibers and mineralization of blood vessels with focal aggregates of lymphocytes within the lamina. CHF was diagnosed in a mole king snake (*Lampropeltis calligaster rhombomaculata*) with myocardial collagen proliferation and osteoid-like material. The etiology of this disorder was unclear; however, microscopic examination of this snake's kidneys and liver revealed severe heterophilic inflammation. Cardiomyopathy of undetermined origin was also reported in two pythons: a Children's python (*Liasis childreni*) and in a juvenile Burmese python (*Python molurus bivittatus*).[2]

Restrictive cardiomyopathy with alteration of the myocardial compliance and CHF was diagnosed for the first time in a 2-year-old captive lethargic Carpet python (*Morelia spilota variegata*), based on echocardiography, Doppler and histopathology. The disease was characterized by left atrial dilation, displacement of the interatrial septum, subsequent right atrial collapsus and sinus venosus dilation with thrombosis in the sinus venosus.[18]

Myocardial degeneration can occur as a result of metabolic disorders and can be observed grossly. For example, gout (of metabolic, nutritional or iatrogenic origin) can cause deposition of urate crystals in multiple tissues including the myocardium. With vitamin E and selenium deficiencies, the myocardium undergoes degeneration with loss of myocytes replaced by fibrous stroma and gross whitish/grey discoloration.[2]

The myocardium and the major arteries may also be subject to mineralization (calcification) when a diet contains excessive levels of calcium and vitamin D. Aortic rupture associated with mineralization of the aorta attributed to secondary nutritional hyperparathyroidism is reported in lizards.[2]

Non-infectious valvulopathies (valvular endocardiosis)

There is no current evidence for primary degenerative endocardiosis occurring in reptiles. However, occasionally reported cardiac conditions of undetermined etiology suggest that this might be a possibility. For example, right atrioventricular insufficiency resulting in bilateral heart failure was observed in a young carpet python (*Morelia spilota variegata*).[19] The lack of septation in squamates may cause sufficient elevation in diastolic pressure to cause bilateral heart failure. Valvular regurgitation can be confirmed with echocardiography.

Pericardial diseases

Pericardial effusion can be inferred from radiography but can only be confirmed by ultrasonography. It can also be an unexpected finding during necropsy (Figure 4).

Visceral gout can lead to thickening of the pericardium with urate crystals deposition, which is easily recognizable by endoscopy or during post-mortem examination.84 Traumatic hemopericardium with the formation of hematomas can occur as a result of cardiac blood sampling in snakes. It has also been reported in a green iguana (*Iguana iguana*) in conjunction with a myocardial abscess.

Pericardial effusion has also been reported in tortoises. A Californian desert tortoise (*Xerobates agassizi*) with cervical fractures and osteomyelitis also had pericardial effusion but no attempt was made to investigate the implication and underlying cause.[20] Echocardiography, radiography and necropsy of an 80-year-old male spur-thighed tortoise (*Testudo graeca*) with post-hibernation anorexia and lethargy revealed pericardial effusion, atrial dilatation, six liver lesions containing choleliths, pneumonia and/or pulmonary edema. These findings were confirmed by post-mortem after non-response to treatment.[21] Pericardial effusion has also been reported in a 2-year-old male bearded dragon (*Pogona vitticeps*) suffering from atherosclerosis in the major arteries of the base of the heart.[22]

Pericardial effusion can also be secondary to neoplasia, as was the case in a female reticulated python (*Malayopyhton reticulatus*) diagnosed with endocardial fibrosarcoma at the base of the aorta. A small amount of fluid within the pericardial sac may be normal and therefore volume measurement is advised during post-mortem.[23]

Neoplasia

Primary neoplasia of the cardiovascular system is uncommon in reptiles. Cardiac tumors in snakes are frequently visible as a distinct mass located approximately at the level of the heart. Reported cases include a cardiac rhabdomyosarcoma in a boa constrictor (*Boa constrictor*) and a black kingsnake (*Lampropeltis getula nigrita*), a cardiac hemangioma of the left atrium in a corn snake (*Pantherophis guttatus*), a cardiac hemangiosarcoma in a Copperhead (*Agkistrodon contortrix*) and a Madagascar giant hognose snake (*Leioheterodon madagascariensis*), a fibrosarcoma in a Gaboon viper (*Bitis gabonica*) and an endocardial fibrosarcoma in a reticulated python (*Malayopython reticulatus*).[2]

Metastasis to the heart has also been reported: metastatic chondrosarcoma and oviductal adenocarcinoma have been reported in a corn snake with metastasis to the heart. Similarly, a disseminated mast cell tumor in an eastern king snake (*Lampropeltis getulus getulus*), multicentric lymphoblastic lymphoma in a loggerhead sea turtle (*Caretta caretta*) and lymphoblastic malignant lymphoma in a boa constrictor (*Boa constrictor*) affected the heart as well as other organs.

Disseminated coelomic papillomas affecting various organs, including the heart, were found in 39% of stranded green turtles (*Chelonia mydas, n=255*) with fibropapillomatosis.[2]

Infectious cardiac disease

Bacterial (endocarditis, myocarditis and pericarditis)

In captive reptiles, infections of the cardiovascular system are often secondary to systemic infections. Endocarditis is most commonly associated with aerobic gram negative infections. *Salmonella* spp., for example, has been isolated in various species of snakes: ultrasonography and angiography revealed a blood flow obstruction in the right atrium of a Burmese python (*Python molurus bivittatus*); post-mortem isolation confirmed *S. arizonae* in addition to *Corynebacterium* spp.[24] *S. arizonae* was also isolated in a coelomic culture in a Dumeril's Boa (*Acrantophis dumerili*).[25] This was concomitant with a granulomatous myocarditis associated with a fibrinous and necrotic pericarditis (Figure 5), as well as hepatitis, pneumonia and thyroiditis with bacterial infiltration. *Salmonella enterica* was also diagnosed in a Burmese python (*Python molurus bivittatus*) with valvular endocarditis and dilated pulmonary trunk and sepsis. Pericarditis can also be associated with fibrinonecrotic lesions.[26]

Concomitant bacterial pneumonia and valvular insufficiencies suggesting bacterial endocarditis can be observed (but without a demonstrated cause-effect relationship): sino-atrial and atrioventricular insufficiencies were diagnosed on echocardiographic examination in a Burmese python (*Python molurus bivittatus*) suffering from concurrent pneumonia caused by *Pseudomonas aeruginosa*.[27] Histopathological examination of the valve was not performed as the snake survived. An Argentine boa (*Boa constrictor occidentalis*) presenting with cardiomegaly, pneumonia and inclusion body disease, showed a large vegetative lesion on the right atrioventricular valve with valvular insufficiency, a mildly dilated right atrium, and pulmonary hypertension on echocardiographic examination. Cultures of a tracheal wash sample resulted in growth of *Ochrobactrum intermedium* and *Pseudomonas putida*. Postmortem examination confirmed the presence of pneumonia and bacterial endocarditis with dystrophic mineralization of the right atrioventricular valve, with different bacteria (gram-positive) than those cultured from the tracheal wash.[28]

Flavobacterium meningosepticum and *Vibrio damsel* have been isolated from the myocardium of a Barber's map turtle (*Graptemys barbouri*) and a vegetative thrombus attached to the left atrioventricular valve and interventricular wall of a stranded leatherback turtle (*Dermochelys coriacea*). *Mycobacterium* spp. was reported in granulomatous myocarditis in a frilled lizard (*Chlamydosaurus kingii*). *Chlamydophila* spp. has also been associated with cardiovascular disease. Reported cases include granulomatous pericarditis and myocarditis in a puff adder (*Bitis arietans*), necrotizing myocarditis in green sea turtles (*Chelonia mydas*) and histiocytic granulomas (determined by transmission electron microscopy) in the heart and organs of an emerald tree boa (*Corallus caninus*).

Mycoplasma alligatoris. is responsible for pericarditis (and for pneumonia and arthritis) in American alligators (*Alligator mississipiensis*).[2]

A diagnosis can be obtained with ultrasound, hemoculture, postmortem tissue culture and/or histopathological examination (valves, myocardium, pericardium). In cases where bacterial infection in confirmed, antimicrobial treatment can be attempted.

Viral (myocarditis, degeneration, necrosis).

Inclusion Body Disease (IBD), thought to be caused by a reptarenavirus, affects snakes and causes large eosinophilic intracytoplasmic inclusion bodies in many different organs including the heart. Acute death of a Savannah monitor (*Varanus exanthematicus*) with myocarditis, and Rosy boas (*Lichanura trivirgata*) with hepatitis and endocarditis were attributed to an adenovirus. Flaviruses have been reported to cause myocardial degeneration and necrosis in farmed American alligators (*Alligator mississippiensis*).[7]

Parasitic

Encapsulated cestodes have been reported within the heart of lizards, with one study finding a prevalence of 5% in wild whiptail lizards (*Cnemidophorus* spp.; *n*=220). Filarial nematodes within the cardiovascular system can cause edema, thrombosis and necrosis. *Macdonaldius oschei* has been observed in fresh blood smears and myocardial histology in various species of reptiles. Trematodes have been reported in the heart chambers and major vessels of chelonians (attached to, or found freely floating within the lumen) in wild and wild-caught chelonians. Green sea turtles (*Chelonia mydas*) can be infested with spirorchid flukes that cause arteritis, endocarditis, and thrombosis, and aneurysms. Similarly, spirorchid eggs and adult *Learedius learei* were observed inside the heart of black sea turtles (*Chelonia mydas agassizii*). Spirorchid eggs were observed in multiple organs and associated with granulomatous lesions in the myocardium of *Trachemys scripta elegans* and *Chrysemys picta*.89 Besnoitiosis and sarcosporidiosis have also been observed in histological sections of ophidian hearts.98 All these parasitic diseases were diagnosed post-mortem. The use of targeted anthelminthic drugs may be a treatment option when ante-mortem parasite identification is performed (fecal examination, blood smear cytology).[2]

DIAGNOSIS

Auscultation, electrocardiography, radiography, and ultrasound are the most used tools for investigating cardiac disease in reptiles. Cardiopulmonary function is strongly influenced by temperature, stress, and the method of physical restraint in these animals. Clinicians should therefore ensure that they are conducting a physical examination once the reptile is calm and within its preferred optimal temperature (which can be measured using an infrared thermometer) and always conduct a full examination to evaluate for concurrent disease and determine if cardiac disease is primary or secondary. The method of restraint should be consistent across patients of a same species and mimic their natural resting position, or the position in which cardiac parameters were established for that species.

Physical examination

Clinical signs of cardiovascular diseases in reptiles include cardiomegaly (evidenced by swelling around the cardiac region) (Figure 6), cyanosis, peripheral edema (e.g. around the mandibular region), pulmonary edema, and ascites. Decreased cardiac output can result in exercise intolerance, syncope, tachycardia, cyanosis, decreased peripheral perfusion, and arrhythmias. Ataxia and head tilt may be observed in the case of brain anoxia secondary to heart failure, atherosclerosis, or arteriosclerosis.[2,18]

Congestive heart failure (CHF) is the inability of cardiac muscle to maintain adequate cardiac output. CHF results in chronic, excessive compensatory mechanisms to maintain cardiac output and blood pressure. The pathophysiology of CHF is probably similar across vertebrate taxa, with some differences between mammals and reptiles, mainly in terms of anatomical structure, physiological regulation, and the relative influences of neural (autonomic) *versus* humoral (kallikrein-kinin, renin-angiotensine, endothelin) systems. Non-crocodilian reptiles possess a single ventricle with a different mechanism of blood ejection, which allows them to shunt blood from right to left (R-L) under certain natural or pathological conditions. These features influence the pathophysiology of heart failure, notably edema formation.[1,2,9,10,16,17] They can also be greatly affected by cardiac diseases with implications for blood oxygenation and ventilation/perfusion mismatch. In addition, reptiles (especially lizards) possess salt glands that act as additional osmoregulatory organs to the renal and gastrointestinal systems.

Clinical signs of CHF include pulmonary edema and coelomic effusion (due to increased preload) and cardiac remodeling, such as cardiomegaly (due to increased afterload). Pitting edema is uncommon. Unlike in mammals, coughing is not a feature of CHF because reptiles lack a true diaphragm.

Auscultation

Heart sounds in reptiles are muffled and can be difficult to auscultate with a standard stethoscope. A digital stethoscope (e.g. Thinklabs One®-Thinklabs, 6500 South Quebec, Suite 210, Centennial CO 80111) or a pressure sensitive acoustic stethoscope (e.g. Ultrascope®-Parker Medical Associates LLC, aka Ultrascope Inc., 2401-C Distribution street, Charlotte NC 28203) can be used to diagnose heart murmurs, especially in large snakes (e.g. boids) and in large lizard species where the heart lies in the mid-coelomic region (varanids), as opposed to within a bony pectoral girdle (e.g. agamid lizards, chameleons, iguanids, lacertids, skinks).

A continuous Doppler ultrasonic probe (8 MHz) can be used as an adjunct to cardiac auscultation to measure heart rate (HR) and estimate cardiac rhythm in all species. The probe is placed with an acoustic coupling gel at the level of the heart or large efferent arteries.

Blood pressure measurement and pulse oximetry
As in mammals, the autonomic nervous system in reptiles regulates blood pressure: the parasympathetic outflow reduces blood pressure *via* the acetylcholine neurotransmitter, while the sympathetic outflow increases blood pressure *via* adrenaline and noradrenaline neurotransmitters.[29] The oculocardiac reflex can be exploited as an effective short-term method of restraint for large non-cooperative lizards (e.g. iguanids and monitors). This method involves applying gentle pressure on the closed eyelids applied for 30 seconds to 2 minutes to stimulate the vagal nerve and thereby reduce heart rate and blood pressure (Fig 9).6 In mammals, two additional antagonists hormonal systems mediate blood pressure: the kallikrein-kinin system (hypotensive) and the renin-angiotensin system (hypertensive). The importance of these systems in non-mammalian vertebrates is poorly understood. In snakes, the bradykinin ligand is structurally different from its mammalian counterpart and its effects on cardiovascular function is species-dependant.[29] The physiology of the renin-angiotensin system has been described in some reptile species, indicating that angiotensine (AT) receptors cannot be classified as AT1 or AT2 receptors seen in mammals. Blood pressure regulation is not necessarily dependent on body temperature; for example constant blood pressure in slider turtles *Trachemys scripta*) can be observed during warming and cooling phases, indicating barostatic regulation due to hysteresis of HR as body temperature changed.[30]

Blood pressure can be measured either directly or indirectly. Direct measurement is achieved through catheterization of the carotid or femoral artery and requires sophisticated equipment, however it produces the most reliable measurement. Indirect measurements are more suited for general practice, as they can be performed using a sphygmomanometer on the forelimb or tail. In medium- and large-sized turtles and lizards, the cuff can be placed on the proximal front limb with the sensor above the brachial artery, in the palmar region of the radius-ulna, or at the proximal level of the tail in large specimens. In snakes, the cuff is placed caudal to the cloaca (coccygeal artery).[31]

Significant differences between invasive and non-invasive blood pressure measurements have been observed in green iguanas (*Iguana iguana*) (Hernandez-Divers, unpublished data, 2007).[31] Comparative studies between both techniques are therefore warranted and results obtained using sphygmomanometers should be interpreted with caution. For example, in three boid species (boa constrictors, carpet pythons, and reticulated pythons), an oscillometric unit (MDE Escort Multiparameter Monitor®, Medical Data Electronics, Arleta, CA.) overestimated systolic arterial blood pressure, and underestimated diastolic and mean arterial blood pressure.[32] Nevertheless, blood pressure range using non-invasive techniques have been described across many species. Chelonians have the lowest mean arterial pressure (15–40 mmHg), while in some lizards (e.g. chameleons) it can be closer to that of mammals (60–80 mmHg). In green iguanas, mean arterial pressure is 40–50 mmHg; under 1% isoflurane anesthesia, the average systolic pressure is 43 (+/- 7) mmHg and average diastolic pressure is 29 (+/- 4) mmHg.[31] In snakes, indirect systolic blood pressures have been reported between 40 and 90mmHg and the average blood pressure increases in proportion to animal length.[31] In arboreal snakes, blood pressure increases in proportion to the distance between the heart and the head; this relation is less pronounced in terrestrial snakes (e.g. *Python bivittatus*), as they are less affected by variations in gravitational force.[33]

Reptilian hemoglobin generally has lower oxygen affinities compared to that of mammals and reptilian red blood cells can deliver oxygen to tissues even at low blood oxygen levels. In theory, pulse oximetry can be used to monitor oxygen saturation levels in response to treatment; however, its use has not been validated in reptiles.[34] Results should always be interpreted with caution because pulse oximeters are designed and calibrated for use in mammals, which have a different hemoglobin structure and oxygen hemoglobin dissociation curve compared to reptiles. In green iguanas, oxygen saturation is lower in pulse-oximetry compared to arterial blood gas analysis.[34] Pulse oximetry can be used to measure relative change in oxygen saturation during treatment, although care should be taken to maintain a constant environmental temperature, as this will influence oxygen saturation. The sensor can be placed in the cloaca or esophagus. Similarly, capnography is not validated for reptiles, and results are particularly difficult to interpret in species with cardiac R-L shunting where mixing of deoxygenated and oxygenated blood occurs.

Clinical Pathology
Blood levels of creatine kinase (CK) and lactate dehydrogenase (LDH) can increase following damage to cardiac muscle, such as myocarditis or infarcts.[35] This must be differentiated from other causes, as increased plasma CK can also result from skeletal muscle damage (e.g. traumatic injuries), intramuscular injections of caustic drugs (e.g. enrofloxacin), and loss of body condition. In mammals, cardiac troponins (troponin I and troponin T) are regulatory proteins that control calcium mediated interaction between actin and myosin; both can be used to measure detect myocardial damage. Unfortunately, the use of cardiac troponins to measure myocardial damage in reptiles has not been evaluated, although they might be of use to assess progression of cardiac dysfunction.

Several lactate dehydrogenase isoenzymes are described in reptiles, although their diagnostic utility has not yet been described. Dyslipidemias are known to occur in reptiles, such as hypercholesterolemia associated with atherosclerosis in a bearded dragon.[22] Hypocalcemia—commonly seen in cases of nutritional disorders—can affect striated cardiac muscle and be visualized as electrocardiographic abnormalities.[2,35] A complete blood count and blood smear can be highly valuable in certain cases; leukocytosis, lymphocytosis, and heterophilia can indicate infections (e.g. vegetative or septic endocarditis) and abnormal red or white blood cells can lead to a diagnosis of hematopoietic neoplasia.[36]

Electrocardiography
General principles
The application and interpretation of electrocardiography (ECG) in reptiles follows the same principle as in mammals. The main challenge associated with ECG in reptiles is related to the low electric amplitudes (typically <1.0mV) that frequently generate electrocardiograms of poor diagnostic value. Standard reference ranges are only available for some species, including red-eared sliders (*Trachemys scripta elegans*)[37], green iguanas (*Iguana iguana*)[38], Moreau's tropical house gecko (*Hemidactylus mabouia*)[39], Gomeran giant lizard (*Gallotia bravoana*)[40], bearded dragons (*Pogona vitticeps*)[14], Russell's vipers (*Daboia russelii*)[41], Indian spectacled cobras (*Naja naja*)[41], constrictor boas (*Boa constrictor imperator*), green anacondas (*Eunectes murinus*), California king snakes (*Lampropeltis getulus californiae*), rat snakes (*Pantherophis obsoletus*)[42], corn snakes (*Pantherophis guttatus*)[43], ball pythons (*Python regius*)[38], and one species of crocodilian; American alligators (*Alligator mississipiensis*).[44] ECG readings are also available for several species in the order squamata but were obtained in the 1960s and might therefore not be accurate.[45]
ECG readings are influenced by lead placement, environmental temperature, and stress. It is therefore of critical importance that readings are taken in the same conditions as when the reference range was established. ECG reading should ideally be recorded when the reptile is calm, in its preferred optimal temperature zone, and with a constant heart rate.

Electrode placement
Correct electrode placement is essential to avoid erroneous ECG readings. The electrodes can be attached to the reptile using self-adhering skin electrodes or alligator clips.[37-45] Placement varies according to species, as described below.

In snakes, the cranial limb leads (negative yellow and positive red leads) are placed approximately two heart lengths cranial to the heart, slightly to the left side at the level of the junction between the dorsal and ventral scales. The caudal limb leads (neutral green and positive black leads) are also placed two heart lengths caudal to the heart, but on the right lateral side.[41] This provides the equivalent tracings of lead II in mammals. A second placement scheme is possible by placing the cranial leads cranial to the heart on each side of the body (red lead on the right side and yellow lead on the left side), while the caudal leads are placed caudal to the heart (green lead on the left side and black lead on the right side).[46] A third option includes following Einthoven's triangle: the yellow lead is placed one heart length cranial to the heart on the right side, and the red lead is placed 60–75% of the snout-to-vent distance from the head, on the left side. The green lead is placed on the right side, across from the positive lead (Figure 7).[47,48]

In chelonians, the two red and yellow cranial electrodes can be placed in the cervical region, lateral to the neck, and the green and black caudal leads are placed on the skin of the knees or caudally to the pelvic limbs (Figure 8).[37,38,45,49] In a study involving a cohort of 72 turtles and tortoises (representing 20 species), adhesive electrodes were placed on the plastron and clamp electrodes to the skin folds. ECG waves could be detected in 41 (57%) animals and complete PQRST complexes were observed in only 19 (26%) animals.[49]

In lizard species where the heart is situated within the pectoral girdle, the cranial electrodes are placed on the lateral cervical region (red lead on the right, yellow lead on the left) and the caudal electrodes are placed on the lateral body wall just caudal to the elbow (green lead on the left, black on the right) (Figure 9A).[14] The caudal electrodes can also be placed on the lateral body wall just cranial to the stifle, with the stifle flexed against the body wall (Figure 9B).[14]

In lizard species where the heart is caudal to the pectoral girdle (i.e. varanids), the traditional four-limb placement can be used (i.e. one lead on each limb, proximally up to the elbows and knees).[45,46] The same configuration applies to crocodilians (Figure 10).[45,46]

Interpretation
The interpretation of ECG readings in reptiles follows the same principles as that in mammals, with analogous P, QRS, and T complexes (Figure 11). Compared to similar-sized mammals, reptile ECGs display P wave pleomorphism, substantially reduced Q and S deflections, and prolonged Q-T intervals.[37-45] In a review of ECG recordings in three reptile species (green iguanas, N=4; ball pythons, N=4; and American alligators, N=4) kept at 35–37°C, heart rate was on average 4-fold lower, atrial and ventricular conduction (longer P- and QRS-wave durations) 2-fold slower, and PR intervals and QT intervals 4-fold longer compared to that of mammals and birds.[38]

Findings specific to reptiles include the SV wave, which represents the depolarization of the *sinus venosus* and the caudal *vena cava*. It is preceded by the T wave (ventricular repolarization) and followed by the P wave (atrial depolarization). The SV wave is of low amplitude (typically ≤ 0.1mV) and can be difficult to separate from the T wave. It can also be obscured by background skeletal muscle activity. The mean electrical axis is difficult to determine due to low voltage potential and is of limited value.[2]

Widened QRS complexes and prolonged Q-T intervals have been observed in a green iguana with aortic stenosis and atrioventricular dilatation.[50] Tall and wide QRS complexes were reported in a carpet python (*Morelia spilota variegata*) with atrioventricular valve insufficiency. Increased P-R intervals associated with decreased Q-T intervals are described in a carpet python with atrioventricular valve insufficiency.[19] A first-degree atrioventricular block was diagnosed in a bearded dragon with hypertensive heart disease and severe atherosclerosis.[52]

Diagnostic imaging
Radiography
Radiography is of considerable diagnostic value in general herpetological practice, but of limited use when it comes to reptile cardiology. This is due to anatomical features that limit x-ray penetration and the absence of standard reference parameters.[53]

Radiography is of most use in snakes, where the position of the lungs around the heart provides sufficient contrast to evaluate the shape and size of the cardiac silhouette (Figure 12A,B).[54] The potential for normal physiological variations in heart size should be taken into consideration when assessing cardiac enlargement. In Burmese pythons (*Python bivittatus*), ventricular myocardium can

increase by up to 40% within 48 hours of feeding as a result of increased metabolic demands and subsequent oxygen consumption.
The heart cannot be visualized in chelonians and crocodilians because it is superimposed by other visceral organs and is masked by osteoderms.[53] The utility of cardiac radiography is also limited in lizard species where the heart is obscured by the bony pectoral girdle.[55] Computed radiography and direct digital radiography can overcome some of these limitations through digital enhancement such as enhancing sharpness, contrast, brightness, and application of filters to enhance contours. The absence of reference parameters can be overcome to a certain extent by performing and comparing radiographs during routine health evaluations.

As with any radiographic examination, a dorsoventral and a horizontal lateral view should be captured. In chelonians, a craniocaudal view can be useful, but this is mostly indicated for visualizing lung fields.[56]

Great vessels should always be assessed in addition to the heart, as mineralization can be observed secondary to hypervitaminosis D3 or other metabolic disturbances.

Echocardiography

General principles

Ultrasound is a safe and non-invasive tool of great diagnostic utility in reptile cardiology, especially when paired with Doppler examination. However, echocardiography can be challenging in crocodilians, chelonians, and many species of lizards, due to the presence of osseous structures (e.g. pectoral girdle, plastron, osteoderms) that impede ultrasound penetration.

Echocardiography has substantially contributed to current knowledge on reptile cardiac anatomy and physiology. These include the description of cardiac circulation in ball pythons (*Python regius*)[10], savannah monitors (*Varanus exanthematicus*)[9], and American alligators (*Alligator mississipiensis*).[15] Equally, B-mode, color-Doppler, and contrast-echocardiography have led to the description of cardiac anatomy and intracardiac blood flow in Hermann's tortoises (*Testudo hermanni*) and Russian tortoises (*Testudo horsfieldi*).[58]

Standard reference parameters for echocardiographic evaluations are being progressively established; standardizations and reference ranges are available for Burmese pythons (*Python bivittatus*)[59], ball pythons (*Python regius*)[60,61], red-tailed boas (*Boa c. constrictor*)[62], bearded dragons (*Pogona vitticeps*)[63], green iguanas (*Iguana iguana*)[64], red-eared sliders (*Trachemys scripta elegans*)[65], Hermann's tortoises (*Testudo hermanni*)[58], Russian tortoises (*Testudo hosfieldi*)[58], and giant Aldabra tortoises (*Aldabrachelys gigantea*)[66]. Typical cardiac parameters include longitudinal and transverse ventricular and atrial lengths measured during diastole and systole, transverse and longitudinal fractional shortenings, diastolic and systolic areas and volumes, and ventricular wall thicknesses and velocities.

Most ultrasound machines used in small animal practice are suitable for reptiles. A generous application of acoustic coupling gel should be applied a few minutes prior to the examination to remove air pockets between the scales. Sedation is rarely required; keeping the eyes covered with a cloth will generally reduce stress and unexpected movement. Selecting ultrasound frequency depends on the need for spatial resolution and imaging depth: lower frequencies produce less resolution but penetrate deeper into the body. Thus, these considerations dictate the frequency of the transducer probes relative to the weight of the patient: 15–20 MHz or greater transducer (reptiles weighing 250g or less), 10–15 MHz (250g–2kg), 5–8 MHz (2–20 kg), and 3–5 MHz (>20kg). If available, an epicardiac transducer can also be used to obtain high-resolution images of the heart. Linear arrays are often used in snakes and lizards, while biconvex probes are recommended for chelonians due to the narrow acoustic windows available (cervicobrachial and prefemoral fossas).[2,67]

Snakes

In boids, three views are recommended for a complete cardiac evaluation.[59] First, a ventral approach involves placing the probe ventral to the heart to view the organ from the caudal ventricular apex to the cranial atria, the *sinus venosus*, the atrioventricular junctions, and the three arterial trunks (Figure 13A). A first short-axis transapical section shows the outline of the apical myocardium and the pericardium as an echogenic line. By moving the probe cranially, a transventricular subarterial short-axis section allows visualization of the three ventricular sub-chambers (the *cavum venosum*, the *cavum arteriosum,* and the *cavum pulmonale*), the interventricular (vertical) septum located between the *cavum arteriosum* and the *cavum venosum*, and the muscular ridge (horizontal septum) marking the separation between the *cavum venosum* and the *cavum pulmonale*. By moving the probe cranially, the resulting transarterial short-axis section shows the two aortic arches of equal diameter and the pulmonary trunk of larger diameter (known as the "Mickey-Mouse-head" section). By slightly moving the probe towards the right, a transatrial short-axis section can be obtained, showing the opening of the *sinus venosus* into the right atrium and both sinoatrial valve. Long-axis sections are visualized by rotating the probe 90° in relation to the previous position. These show both atrial cavities opening into the single ventricle. The left atrioventricular junction can be observed by orientating the ultrasound plane ventrodorsally from the right to the left. The right atrioventricular junction can be observed by orientating the ultrasound plane ventrodorsally from left to right. The monucuspid atrioventricular valves can also be visualized in this view (Figure 14).[59]
The second (right intercostal) and third (left intercostal) involve placing the probe laterally on either side of the body to view the three arterial trunks, both atria, and the single ventricle (Figure 13B).[59] Reference values are provided for the ball python (*Python regius*).[60]

Chelonians
In chelonians, the optimal view is obtained through the right cervical window (Figure 15A). In soft-shelled turtles (e.g. *Trionyx* spp., *Apalone* spp., *Cyclanorbis* spp.) and pancake tortoises (*Malachochersus tornieri*), as well as in neonate or juvenile chelonians, the probe can also be pressed directly onto the cartilaginous plastron, which allows for limited ultrasound wave penetration.[2]
The right cervical window approach enables visualization of base to apex inflow and outflow (Figure 15B). Fractional shortening can be calculated by measuring ventricular size, wall thickness, and outflow tracts. Pulsed-wave Doppler is used to record diastolic biphasic atrioventricular flow and the systolic ventricular outflow. This enables calculation of early diastolic (E) and late diastolic (A) wave peak velocities, E/A ratio, ventricular outflow systolic peak and mean velocities and gradients, velocity-time integral, acceleration and deceleration times, and ejection time. Reference values are provided for the red-eared slider (*Trachemys scripta elegans*).[65]

Lizards
In lizard species where the heart is located at the base of the neck, the probe can be positioned cranial or caudal to the bony pectoral girdle, although this window is relatively narrow and can restrict visualization of the entire heart in certain species (e.g. chameleons). Images of better diagnostic quality can be obtained by placing the probe in the left or right axillary region, either in sternal (e.g. iguanids) (Figure 16A) or in lateral recumbency (e.g. agamid lizards).[63,64] In varanids, the heart is located more caudally comparatively to other species and can therefore be easily accessed via a ventral approach, caudal to the sternum and slightly on the right side.[67]

The left and right axillary regions allow for more complete visualization of cardiac structures (Figure 16B). From the left axillary approach, a long axis view shows the atria, the single ventricle, the interatrial septum, and the atrioventricular valve leaflets. In the long axis view, the diameters of both atria can be measured during diastole. The right axillary window allows for the evaluation of pulmonary artery flow. Both right and left views can reveal pericardial effusion or valvular insufficiency. With standardized imaging planes, the diameter of the cardiac chambers, fractional area change, and fractional shortening (in the longitudinal and transverse planes) can be calculated. Reference values are provided for the bearded dragon (*Pogona vitticeps*).[63]

THERAPY

The pharmacokinetics and pharmacodynamics of cardiovascular drugs are poorly understood in reptiles. Unfortunately, reptiles suffering from cardiac disease are often diagnosed *post-mortem* and thus rarely subject to therapeutic trials. Further research is therefore urgently needed to provide evidence-based recommendations for the treatment of cardiac pathologies in reptiles.

Non-pharmacologic interventions
As with mammals, non-pharmacologic management of cardiovascular disease relies on reducing cardiac stress. Interventions include reducing metabolic rates by maintaining the reptile at the lower end of its preferred optimum temperature zone, and reducing stress by minimizing handling and keeping the patient isolated (expect in gregarious species such as Solomon Island skinks [*Corucia zebrata*] and shinglebacks [*Tiliqua rugosa*]). Care should also be given to maintain an adequate diet, as overfeeding has been attributed to death in snakes with cardiomyopathy.[2]

Pharmacologic interventions
In mammals, cardiac therapy is chosen based on the desired effect on heart contractibility and heart rate, vasoconstriction or vasodilatation, or increasing or reducing blood volume. The cornerstone of managing congestive heart failure rests on reducing the volume of extracellular fluid and reducing preload through the use of diuretics and vasodilators. Although extrapolation from mammals to reptiles is hazardous due to significant differences in size, metabolism, anatomy, and physiology, basic principles remain applicable. Oxygen supplementation should be instituted for dyspneic animals, especially in the presence of pulmonary edema. Response to oxygen therapy is, however, difficult to assess in species capable of large right-to-left cardiac shunting when pulmonary vasculature is bypassed (e.g. pythons, varanids, and freshwater turtles).

Inotropes and chronotropes
Inotropes are a group of drugs that alter the contractility of the heart. Positive inotropes such as pimobendan are commonly used in small animal cardiology and increase both survival time and quality of life in the treatment of canine dilated cardiomyopathy. Digoxin is another positive inotrope that increases the cytosolic calcium concentration by inhibiting Na/K ATPase pumps in myocardial cells. Digoxin is also used for its effect in decreasing heart rate (negative chronotrope) and is mainly indicated for atrial fibrillation.

Beta-blockers (e.g. atenolol, propranolol) and calcium channel blockers (e.g. diltiazem) are negative inotropes that are mainly indicated in the treatment of hypertrophic cardiomyopathy in cats and supraventricular arrhythmias. In mammals, their side effects include bradycardia and hypotension. Beta-blockers are uncommonly used for treatment of supraventricular tachycardia and atrial fibrillation in reptile owing to their unknown effectiveness in this taxon. The use of atropine (0.04 mg/kg IV) and glycopyrrolate (0.01 mg/kg IV) to counter the negative chronotropic effects of vagal stimulation of the heart was unsuccessful in the green iguana (*Iguana iguana*).[68] In a study of 15 non-anesthetized healthy juvenile green iguanas, atenolol administered at 1 mg/kg IV significantly reduced resting HR, but did not significantly affect mean arterial blood pressure at the midpoint of the HR range. Failure to alter the baroreceptor reflex function in this instance suggests that atenolol is of little use in regulating blood pressure.[69]

Little evidence exists on the efficacy of cardiac chronotropes in reptiles. In the previously mentioned study of 15 non-anesthetized green iguanas, administration of methylatropine (1 mg/kg IV) increased resting HR but did not significantly affect mean arterial blood pressure.[69]

Angiotensin-converting enzyme (ACE) inhibitors.
In mammals, angiotensin-converting enzyme (ACE) inhibitors block the formation of angiotensin II, which normally promotes venous and arterial vasoconstriction. In addition, ACE inhibitors suppress the production of aldosterone, which normally causes renal sodium and fluid retention. The combined effect of ACE inhibitors therefore reduces preload and afterload through venous and arterial vasodilation, and reduction of blood volume. Enalapril and benazepril are commonly used in dogs and cats due to their longer half-life, but it is unknown if this is also the case in reptiles. In American

alligators, captopril and enalapril have been used to inhibit the pressor response to exogenous angiotensin I (ANG I) and II (ANG II) at a dose of 0.5mcg/kg/h and 300 mcg/kg/h, respectively.[70] Both ACE inhibitors failed to alter baseline blood pressure, significantly reduced the response to ANG I, but not ANG II. When combined with spironolactone and furosemide, enalepril administered at 0.5–0.7 mg/kg q 24 hours per os managed CHF in a spiny tailed monitor (*Varanus acanthurus*) (LA Clayton, CA Hadfield, SR Gore, et al. unpublished data).

Diuretics

Diuretics reduce fluid overload and decrease edema and effusions by promoting water excretion. In mammals, furosemide is the diuretic of choice. It is a loop diuretic that acts by inhibiting sodium, potassium, and chloride co-transporters in the ascending limb of the loop of Henle of the kidney. Although reptiles lack a loop of Henle, furosemide has been successfully used to cause diuresis in chelonians and ophidians.[71-74] The mechanism of action is thought to involve the effect of furosemide on the sodium:potassium pumps located in the urinary bladder or in the terminal rectum (many species of reptiles reflux urine from the cloaca into the rectum to absorb water in conjunction with NaCl). Furosemide should be used with caution in reptile patients with renal disease, because it can cause dehydration and visceral gout. Biochemical parameters including potassium, phosphorus, and major nitrogenous waste product (ammonia, urea, or uric acid) should be monitored and treatment adapted accordingly. Furosemide (two doses at 5.2 mg/kg IM, 3 weeks apart) was successfully used for the treatment of peripheral edema in a spur-thighed tortoise (*Testudo graeca*) with atrial dilatation and pericardial effusion.[75] The peripheral edema resolved within 12 hours after each injection. Treatments with furosemide (2 mg/kg/8-12h SQ for 3 doses) were considered to improve quality of life in two lizards (a 6-year-old bearded dragon and a 13-year-old green iguana) suffering from CHF.[76] However, furosemide (5 mg/kg/d IM) was unsuccessful at managing atrioventricular insufficiency with secondary pulmonary edema and pericardial effusion in a carpet python (*Morelia spilota variegata*).[19]

There is little to no information available on the use and efficacy of other diuretics in reptiles. Hydrochlorothiazide has been used as a diuretic in lizards with renal disease. Methylated xanthines (aminophylline and theophylline) have successfully induced diuresis in different species (*Testudo* sp., *Trachemys scripta elegans*, *Iguana iguana*, *Eublepharis macularius*, and *Pogona vitticeps*). Their diuretic effect is apparently due to increased rates of excretion of sodium and chloride ions by renal tubules, in addition to their positive inotropic effect on cardiac and renal functions (increased renal blood flow and increased glomerular filtration rate).[77] Spironolactone is an aldosterone antagonist and a weak diuretic in mammals that can either be used alone or in conjunction with furosemide to offset the loss of potassium. Nothing is known about its potential use in reptiles.

REFERENCES

1. Farrell AP, Gamperl AK, Francis ETB. Comparative aspects of heart morphology. In: Gans C, Gaunt AS, eds. Biology of the Reptilia, vol. 19, Morphology G, Visceral organs. Ithaca: Society for the Study of Amphibians and Reptiles; 1998: 375-424.
2. Schilliger L, Girling S. Cardiology. In: Divers S, Stahl S, eds. Mader's reptile and amphibian medicine and surgery. St Louis, MO: Elsevier; 2019: 669-698.
3. Lillywhite HB, Albert JS, Sheehy CM, et al. Gravity and the evolution of cardiopulmonary morphology in snakes. Comp Biochem Physiol A Mol Integr Physiol. 2012; 161: 230-242.
4. Funk RS, Bogan JE. Snake taxonomy, anatomy, and physiology. In: Divers S, Stahl S, eds. Mader's reptile and amphibian medicine and surgery. St Louis, MO: Elsevier; 2019: 50-62.
5. Boyer TH, Innis C. Chelonian taxonomy, anatomy, and physiology. In: Divers S, Stahl S, eds. Mader's reptile and amphibian medicine and surgery. St Louis, MO: Elsevier; 2019: 31-49.
6. Barten S, Simpson S. Lizard taxonomy, anatomy, and physiology. In: Divers S, Stahl S, eds. Mader's reptile and amphibian medicine and surgery. St Louis, MO: Elsevier; 2019: 63-74.
7. Nevarez JG, Simpson S. Crocodilian taxonomy, anatomy, and physiology. In: Divers S, Stahl S, eds. Mader's reptile and amphibian medicine and surgery. St Louis, MO: Elsevier; 2019: 75-82.

8. Millard RW, Johansen K. Ventricular outflow dynamics in the lizard, *Varanus niloticus*: responses to hypoxia, hypercarbia and diving. J Exp Biol. 1974; 60: 871-880.
9. Burggren W, Johansen K. Ventricular haemodynamics in the monitor lizard *Varanus exanthematicus*: Pulmonary and Systemic Pressure Separation. J Exp Biol. 1982; 96: 343-354.
10. Starck JM. Functional morphology and patterns of blood flow in the heart of *Python regius*. J Morphol. 2009; 270: 673-687.
11. Jensen B, Boukens BJ, Postma AV, et al. Identifying the evolutionary building blocks of the cardiac conduction system. PLoS One. 2012; 7: e44231.
12. Wang T, Warburton S, Abe A, et al. Vagal control of heart rate and cardiac shunts in reptiles: relation to metabolic state. Exp Physiol. 2001; 86: 777-784.
13. Seymour RS. Scaling of cardiovascular physiology in snakes. Amer Zool. 1987; 27: 97-109.
14. Hunt D. Electrocardiography of the normal inland bearded dragon. Thesis submitted in part fulfilment of the requirements for the Royal College of Veterinary Surgeons (RCVS) Diploma in Zoological Medicine. London, Royal College of Veterinary Surgeons; 2013. 42p.
15. Malvin GM, Hicks JW, Greene ER. Central vascular flow patterns in the alligator *Alligator mississipiensis*. Am J Physiol. 1995: 1133-1139.
16. Hicks JW. Cardiac shunting in reptiles: mechanisms, regulation and physiological functions. In: Gans C, Gaunt AS, eds. Biology of the Reptilia, vol. 19, Morphology G, Visceral organs. Ithaca: Society for the Study of Amphibians and Reptiles; 1998: 425-483.
17. Hicks JW. The physiological and evolutionary significance of cardiovascular shunting patterns in reptiles. News Physiol Sci. 2002; 17: 241-245.
18. Schilliger L, Chetboul V, Damoiseaux C, et al. Congestive heart failure in a McDowell's carpet python (*Morelia spilota mcdowelli*). J Zoo Wildl Med. 2016; 47: 1001-1104.
19. Rishniw M, Carmel BP. Atrioventricular valvular insufficiency and congestive heart failure in a carpet python. Aust Vet J. 1999; 77: 580-583
20. Penninck DG, Stewart JS, Paul-Murphy J, et al. Ultrasonography of the California desert tortoises (*Xerobates agassizi*): anatomy and application. Vet Radiol. 1991; 32: 112-116.
21. Redrobe SP, Scudamore CL. Ultrasonographic diagnosis of pericardial effusion and atrial dilation in a spur-thighed tortoise (*Testudo graeca*). Vet Rec. 2000; 146: 183-185.
22. Schilliger L, Lemberger K, Chai N, et al. Atherosclerosis associated with pericardial effusion in a central bearded dragon (*Pogona vitticeps*). J Vet Diagn Invest. 2010; 22: 789-792.
23. Gumber S, Nevarez JG, Cho DY. Endocardial fibrosarcoma in a reticulated python (*Python reticularis*). J Vet Diagn Invest. 2010; 22: 1013-1016.
24. Jacobson ER. Endocarditis and Congestive Heart Failure in a Burmese Python. J Zoo Wildl Med. 1991; 22: 245-248.
25. Schilliger L, Vanderstylen D, Piétrain J, et al. Granulomatous myocarditis and coelomic effusion due to Salmonella enterica arizonae in a Madagascar Dumerili's boa (*Acrantophis dumerili*, Jan. 1860). J Vet Cardiol. 2003; 5: 43-45.
26. Schroff S, Schmidt V, Kiefer I, et al. Ultrasonographic diagnosis of an endocarditis valvularis in a Burmese python (*Python molurus bivittatus*) with pneumonia. J Zoo Wildl Med. 2010; 41: 721-724.
27. Schilliger L, Trehiou-Sechi E, Petit AM, et al. Double valvular insufficiency in a Burmese python (*Python molurus bivittatus*, Linnaeus, 1758) suffering from concomitant bacterial pneumonia. J Zoo Wildl Med. 2010; 41: 742-744.
28. Wernick MB, Novo-Matos J, Ebling A, et al. Valvulopathy consistent with endocarditis in an Argentine boa (*Boa constrictor occidentalis*). J Zoo Wildl Med. 2015; 46: 124-129.
29. Breno MC, Presoto BC, Borgheresi R, et al. Characteristics of neural and humoral systems involved in the regulation of blood pressure in snakes. Comp Bio and Physiol. 2007; 147: 766-78.
30. Galli G, Taylor EW, Wang T. The cardiovascular responses of the freshwater turtle (*Trachemys scripta*) to warming and cooling. J Exp Biol. 2004; 207: 1471-1478.
31. Martinez-Jimenez D, Hernandez-Divers SJ. Emergency Care of Reptiles. Vet Clin Exot Anim. 2007; 10: 557–585.
32. Chinnadurai SK, Wrenn A, DeVoe RS. Evaluation of noninvasive oscillometric blood pressure monitoring in anesthetized boid snakes. J Am Vet Med Assoc. 2009; 234: 625-630.

33. Enok S, Saly C, Abe AS, et al. Intraspecific scaling of arterial blood pressure in the Burmese python. J Exp Biol. 2014; 217: 2232-2234.
34. Hernandez SM, Schumacher J, Lewis SJ, et al. Selected cardiopulmonary values and baroreceptor reflex in conscious green iguanas (*Iguana iguana*). Am J Vet Res. 2011; 72: 1519-1526.
35. Heatley JJ, Russel KE. Cinical chemistry. In: Divers S, Stahl S, eds. Mader's reptile and amphibian medicine and surgery. St Louis, MO: Elsevier; 2019: 319-332.
36. Schilliger L, Selleri P, Frye FLFL. Lymphoreticular neoplasm and leukemia in a red-tail boa (*Boa constrictor constrictor*) associated with concurrent inclusion body disease. J Vet Diagn Invest. 2011; 23: 159-162.
37. Holz RM, Holz P. Elctrocardiography in anaesthetized red-eared sliders (*Trachemys scripta elegans*). Res Vet Sci. 1995; 58(1): 67-69.
38. Boukens BJD, Kristensen DL, Filogonio R, et al. The electrocardiogram of vertebrates: Evolutionary changes from ectothermy to endothermy. Prog Biophys Mol Biol. 2019; 144: 16-29.
39. Germer CM, Tomaz, JM, Carvalho AF, et al. Electrocardiogram, heart movement and heart rate in the awake gecko (*Hemidactylus mabouia*). J Comp Physiol B 2015; 185: 111-118.
40. Martinez-Silvestre A, Mateo JA, Pether J. Electrocardiographic parameters in the Gomeran giant lizard, *Gallotia bravoana*. J Herp Med Surg 2003; 13(3): 22-25.
41. Karthik M, Rajkumar K, Krishna MK, et al. Electrocardiographic parameters of free living Russel's viper (*Daboia russelii*) and Indian spectacled cobra (Naja naja). AAZV Annual Conference 2013; 174-180.
42. Valentinuzzi ME, Hoff HE, Geddes LA. Observations on the electrical activity of the snake heart. J Electrocardiol. 1969; 2: 39-50.
43. Lewis M, Bouvard J, Eatwell K, et al. Standardisation of electrocardiographic examination in corn snakes (*Pantherophis guttatus*). Vet Rec. 2020; 186(9): 1-8.
44. Heaton-Jones TG, King RR. Characterization of the Electrocardiogram of the American Alligator (*Alligator mississippiensis*). J Zoo and Wildl Med. 1994; 25: 40-47.
45. Mullen RK. Comparative Electrocardiography of the Squamata. Physiol Zool. 1967; 40: 114.
46. Kharin S, Shmakov D. A comparative study of contractility of the heart ventricle in some ectothermic vertebrates. Acta Herpetol. 2009; 4: 57-71.
47. Mitchell MA. Reptile cardiology. Vet Clin North Am Exot Anim Pract. 2009; 12: 65-79.
48. Bogan JE. Ophidian cardiology – A review. J Herp Med Surg. 2017; 27(1-2): 62-77.
49. Nogradi AL, Balogh M. Establishment of methodology for non-invasive electrocardiographic measurements in turtles and tortoises. Acta Vet Hung. 2018; 66(3): 365-75.
50. Clippinger T. Aortic stenosis and atrioventricular dilatation in a green iguana (*Iguana iguana*). In: Proc Am Assoc Zoo Vet; 1993: 342–344.
51. Rishniw M, Carmel BP. Atrioventricular valvular insufficiency and congestive heart failure in a carpet python. Aust Vet J. 1999; 77: 580–583.
52. Schilliger L, Paillusseau C, Gandar F, et al. Hypertensive heart disease and encephalopathy in a central bearded dragon (*Pogona vitticeps*) with severe atherosclerosis and first-degree atrioventricular block. J Zoo Wildl Med. 2019; 50(2): 482-486.
53. Holmes SP, Divers SJ. Radiography-General principles. In: Divers S, Stahl S, eds. Mader's reptile and amphibian medicine and surgery. St Louis, MO: Elsevier; 2019: 486-490.
54. Comolli JR L, Divers SJ S. Radiography-Snakes. In: Divers S, Stahl S, eds. Mader's reptile and amphibian medicine and surgery. St Louis, MO: Elsevier; 2019: 503-513.
55. Holmes SP, Divers SJ. Radiography-Lizards. In: Divers S, Stahl S, eds. Mader's reptile and amphibian medicine and surgery. St Louis, MO: Elsevier; 2019: 491-502.
56. Holmes SP, Divers SJ. Radiography-Chelonians. In: Divers S, Stahl S, eds. Mader's reptile and amphibian medicine and surgery. St Louis, MO: Elsevier; 2019: 514-527.
57. Malvin GM, Hicks JW, Greene ER. Central vascular flow patterns in the alligator *Alligator mississipiensis*. Am J Physiol 1995; R :1133-1139.
58. Prutz M, Hungerbuhler S, Lass M, et al. Contrast echocardiography for analysis of heart anatomy in tortoises. Tierarztl Prax Ausg K Kleintiere Heimtiere. 2015; 43: 231-237.

59. Schilliger L, Tessier D, Pouchelon J-L, et al. Proposed Standardization of the Two-Dimensional Echocardiographic Examination in Snakes. J Herp Med Surg. 2006; 16: 76-87.
60. Paillusseau C, Gandar F, Schilliger L, et al. Two-dimensional echocardiographic measurements in the ball python (*Python regius*). J Zoo Wildl med 2019; 50(4): 976-82.
61. Bagardi M, Bardi E, Manfredi M, et al. Two-dimensional and Dopller echocardiographic evaluation in twenty-one healthy *Python regius*. Vet Med Sci 2021; doi: 10.1002/vms3.426.
62. Eduarda M, Conceicao AM, Monteiro OB, et al. Effect of biometric variables on two-dimensional echocardiographic measurements in the red-tailed boa (*Boa constrictor constrictor*). J Zoo Wildl Med. 2014; 45(3): 672-7.
63. Silverman S, Sanchez-Migallon Guzman D, Stern J, et al. Standardization of the two-dimensional transcoelomic echocardiographic examination in the central bearded dragon (*Pogona vitticeps*). J Vet Cardiol. 2016; 18: 168-178.
64. Gustavsen KA, Saunders AB, Young BD, et al. Echocardiographic and radiographic findings in a cohort of healthy adult iguanas (*Iguana iguana*). J Vet Cardiol. 2014; 16: 185-196.
65. Poser H, Russello G, Zanella A, et al. Two-dimensional and Doppler echocardiographic findings in healthy non-sedated red-eared slider terrapins (*Trachemys scripta elegans*). Vet Res Commun. 2011; 35: 511-520.
66. Campolo M, Oricco S, Cavicchio P, et al. Echocardiographic evaluation of four giant Aldabra tortoises (*Aldabrachelys gigantea*). Vet Rec Open. 2019; 6: 1-6.
67. Hochleitner C, Sharma A. Ultrasonography. In: Divers S, Stahl S, eds. Mader's reptile and amphibian medicine and surgery. St Louis, MO: Elsevier; 2019: 543-549.
68. Pace L, Mader D. Atropine and glycopyrrolate, route of administration and response in the green iguana (*Iguana iguana*). ARAV Annual Conference 2002 ; 79.
69. Hernandez SM, Schumacher J, Lewis SJ, et al. Selected cardiopulmonary values and baroreceptor reflex in conscious green iguanas (*Iguana iguana*). Am J Vet Res. 2011; 72: 1519-1526.
70. Silldorff EP, Stephens GA. Effects of converting enzyme inhibition and alpha receptor blockade on the angiotensin pressor response in the American alligator. Gen Comp Endocrinol. 1992; 87: 134-140.
71. Cipolle MD, Zehr JE. Renin release in turtles: effects of volume depletion and furosemide administration. Am J Physiol 1985; 249: 100-105.
72. Uva B, Vallarino M. Renin-angiotensin system and osmoregulation in the terrestrial chelonian *Testudo hermanni* Gmelin. Comp Biochem Physiol A Comp Physiol 1982; 71: 449-451.
73. Stephens GA, Robertson FM. Renal responses to diuretics in the turtle. J Comp Physiol B 1985; 155: 387-393.
74. LeBrie SJ, Boelcskevy BD. The effect of furosemide on renal function and renin in water snakes. Comp Biochem Physiol C 1979; 63C: 223-228.
75. Redrobe SP, Scudamore CL. Ultrasonographic diagnosis of pericardial effusion and atrial dilation in a spur-thighed tortoise (*Testudo graeca*). Vet Rec 2000; 146: 183-185.
76. Simone-Freilicher E, Sullivan P, Quinn R, et al. Two cases of congestive heart failure in lizards. Exoticscon 2015; 505-509.
77. Girling SJ, Hynes B. Cardiovascular and haemopoietic systems In: Girling SJ,Raiti P, eds. 2nd ed. ed. Cheltenham: Cheltenham : BSAVA - British Small Animal Veterinary Association, 2004; 243-260.

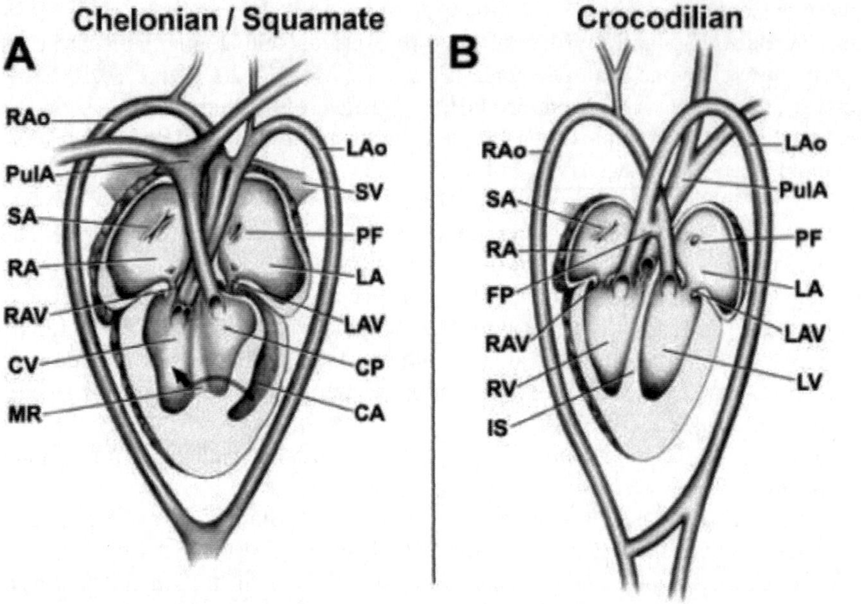

Figure 1: Anatomical drawing of the two reptilian cardiac configurations. **A**, chelonian and squamate pattern; **B**, crocodilian pattern. CA, *cavum arteriosum*; CP, *cavum pulmonale*; CV, *cavum venosum*; FP, foramen of Panizza; IS, interventricular septum; LA, left atrium; Lao, left aorta; LAV, left atrioventricular valve; LV, left ventricle; MR, muscular ridge; PF, pulmonary vein foramen; PulA, pulmonary artery; RA, right atrium; RAo, right aorta; RAV, right atrioventricular valve; RV, right ventricle; SA, sinoatrial valve; SV, sinus venosus. (Modified from Wyneken J, Vet Clin North Am. Exot Anim Pract, 2009, 12: 51-63, with permission from Elsevier).[2]

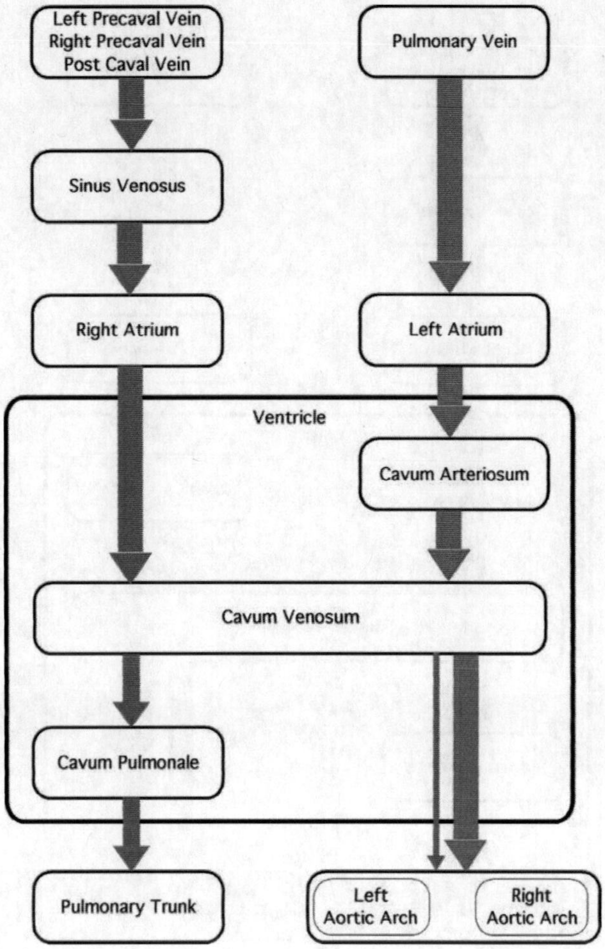

Figure 2: Schematic draw of blood flows during normal breathing in non-crocodilians.

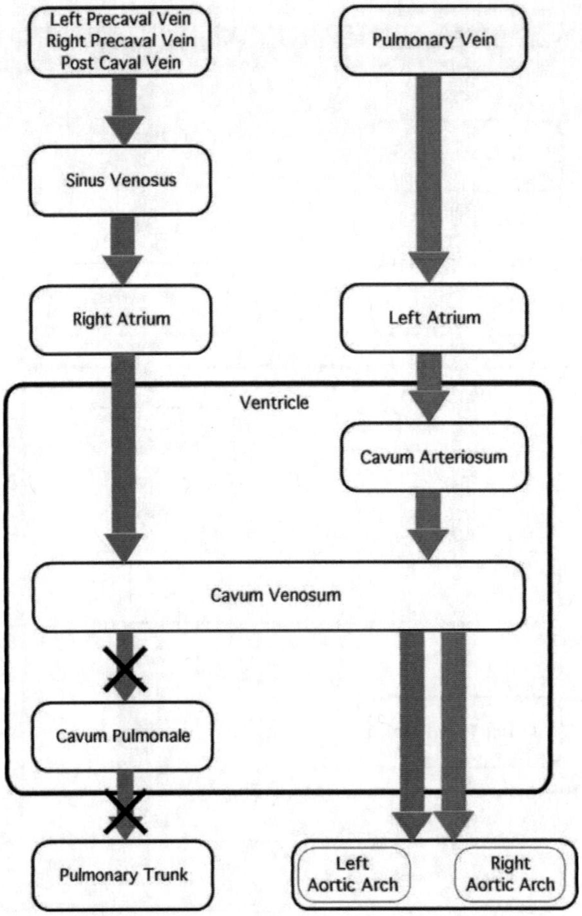

Figure 3: Schematic draw of blood flows during apnea in non-crocodilians.

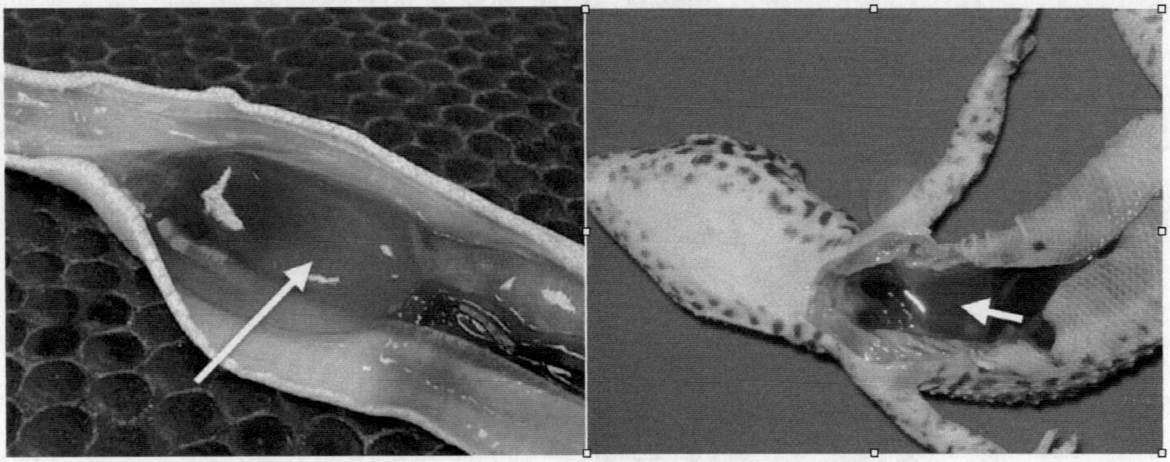

Figure 4: Pericardial effusions (white arrows) in (left) an albino boa (*Boa constrictor*) and (right) a leopard gecko (*Eublepharis macularius*). (Courtesy of S. Girling.)

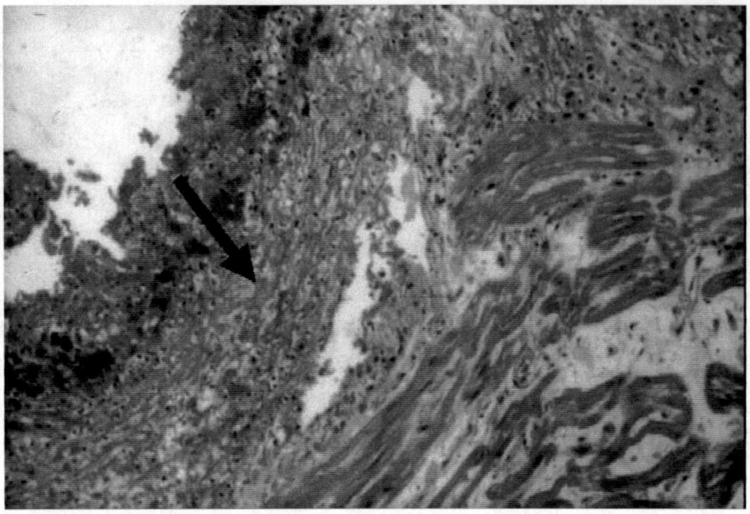

Figure 5: Histological examination of the pericardium, hematoxylin eosin staining (original magnification, x 20) in a Dumleril's boa (*Acrantophis dumerili*). A severe pericarditis (black arrow) is seen.

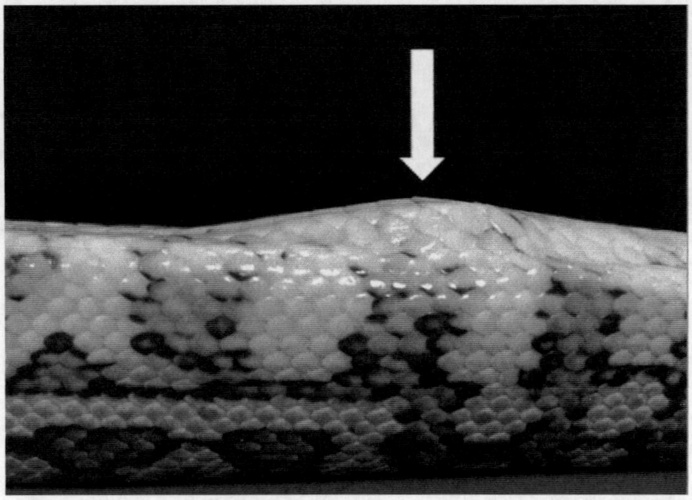

Figure 6: Cardiomegaly (white arrow) in Mc Dowell's Carpet Python (*Morelia spilota McDowelli*).

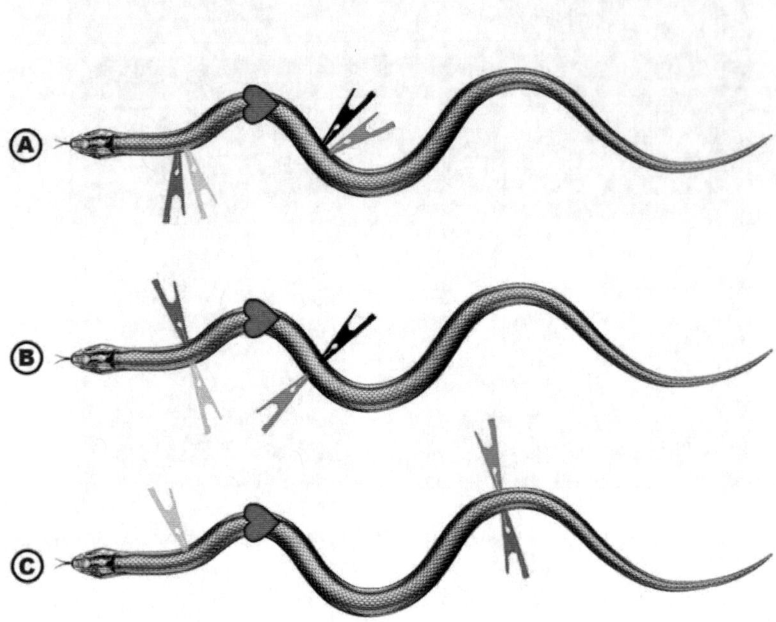

Figure 7: ECG electrode placement in snakes. **A**: according to Karthik M, Rajkumar K, Krishna MK, et al., 2013. **B**: according to Kharin S and Shmakov D, 2009. **C**: according to Mitchell MA, 2009 and Bogan JE, 2017. Figure Pierre Morice.

Figure 8: ECG electrode placement in chelonians according to Holz RM and Holz P, 1995 and Kharin S and Shmakov D, 2009. Figure Pierre Morice.

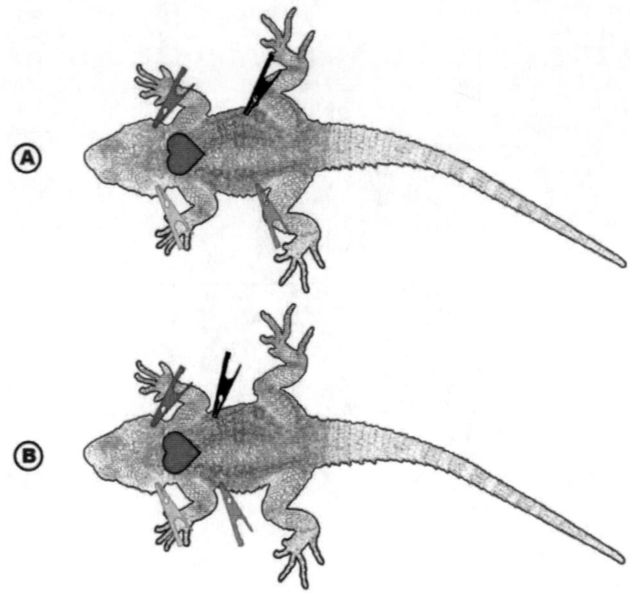

Figure 9: Two options for ECG electrode placement (**A, B**) in non-varanid lizards. According to Hunt, 2013. Figure Pierre Morice.

Figure 10: ECG electrode placement in crocodilians (**A**) and varanids (**B**) according to Mullen RK, 1967 and Kharin S and Shmakov D, 2009. Figure Pierre Morice.

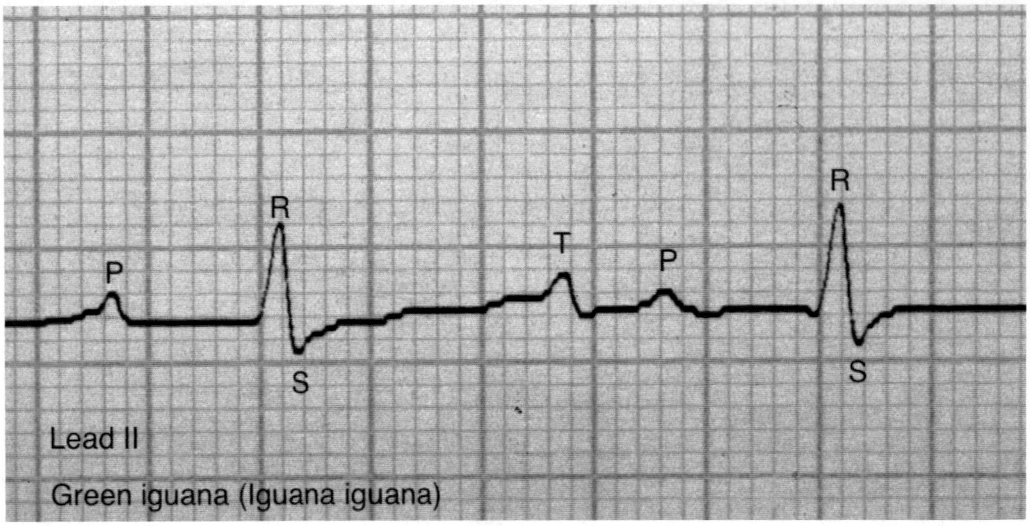

Figure 11: Typical ECG complex in reptiles. SV wave: absent; P: P wave; Q: reduced deflection; R: R wave; T: T wave.

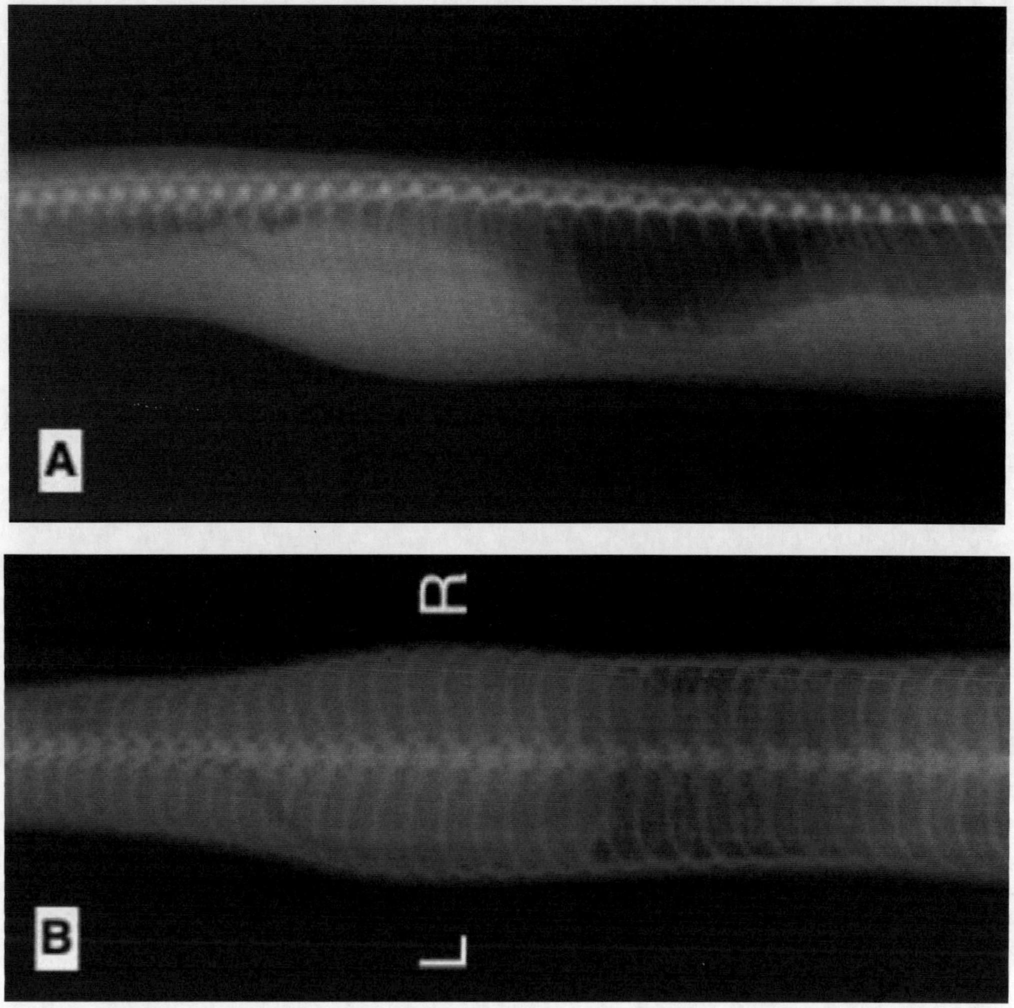

Figure 12: Enlarged cardiac silhouette (cardiomegaly) causing deviation of the trachea towards the left side in a Southern hognose snake (*Heterodon simus*). Latero-lateral (**A**) and dorso-ventral (**B**) incidences.

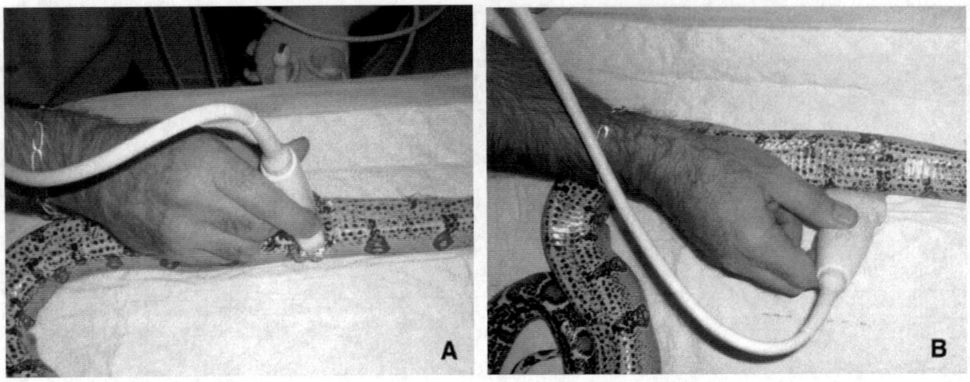

Figure 13: Echocardiographic examination in a true red-tailed boa (*Boa constrictor constrictor*). **A**: ventral approach (short-axis section). **B**: left lateral approach (long-axis sections).

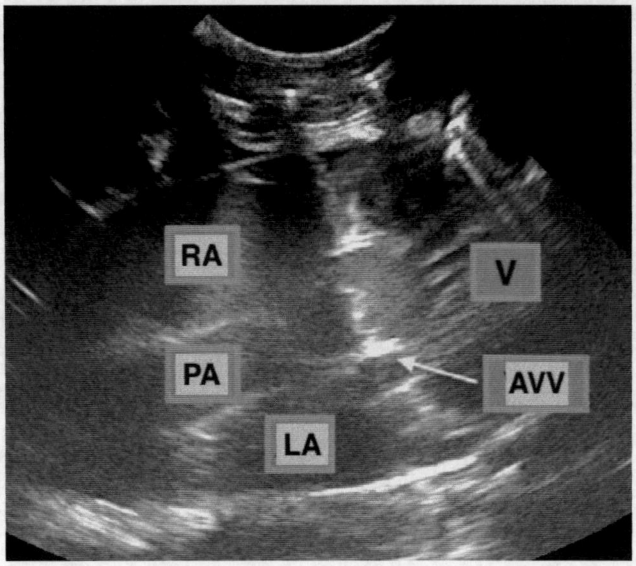

Figure 14: Long-axis section of the left atrioventricular junction in snakes. The monucuspid atrioventricular valves can also be visualized in this view. V: Ventricle; LA: Left Atrium; RA: Right Atrium; AVV: Atrio Ventricular Valve; PA: Pulmonary Artery.

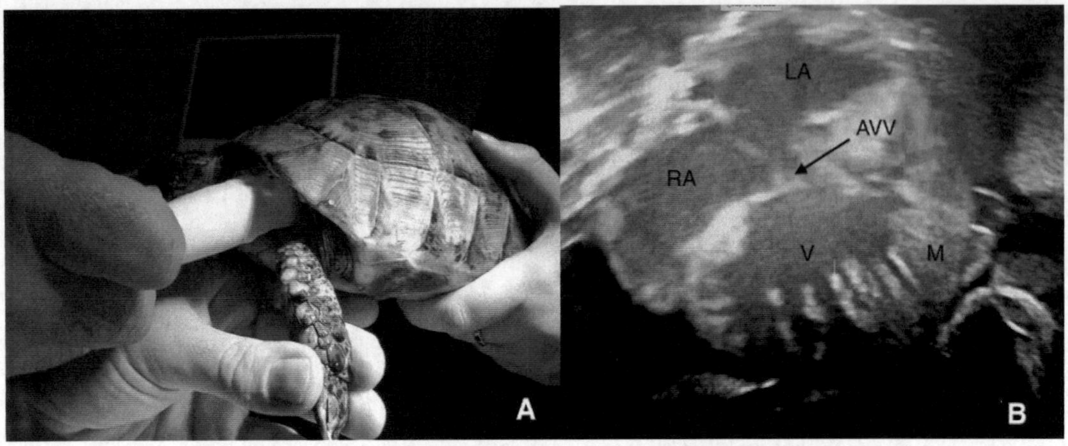

Figure 15: Placement of an ultrasound probe (left cervicobrachial ultrasound window) in a spur-thighed tortoise (*Testudo graeca*) (**A**). Echographic view of the heart in a red-eared slider (*Trachemys scripta elegans*) (**B**).

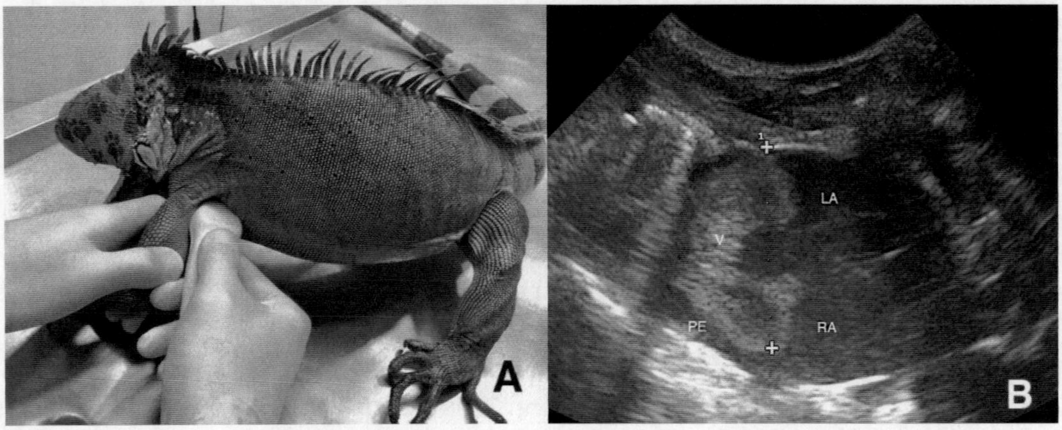

Figure 16: Placement of an ultrasound probe (lateral approach, left axillary ultrasound window) in a green iguana (*Iguana iguana*) in sternal recumbency. Echocardiographic examination in a bearded dragon (*Pogona vitticeps*). V: Ventricle; LA: Left Atrium; RA: Right Atrium; PE: Pericardial Effusion.

Practical Amphibian Medicine

Exoticscon Virtual 2021
Concurrent with AEMV and AAV

Colin McDermott, VMD, CertAqV
Zodiac Pet and Exotic Hospital, Hong Kong

Introduction

Class Amphibia contains approximately 8000 recognized species of frogs and toads (Order Anura), salamanders and newts (Order Caudata), and caecilians (Order Gymnophiona). These species occupy a wide range of habitats and ecological niches on all continents except Antarctica. There are several species of amphibians that are commonly kept as pets, with a wider diversity of species in zoological collections. The recent environmental and disease pressures on wild amphibians have led to an increased focus on conservation, medical management of disease, and improving general care in captivity as part of species survival programs. Amphibian patients are generally a small component of clinical practice; however, the exotics clinician should be familiar with the basic anatomy, physiology, husbandry, and clinical techniques for amphibians. This talk will serve as an introduction to amphibian medicine for the exotics practitioner.

Relevant anatomy/physiology

Metamorphosis

In general, most species of amphibians go through the process of metamorphosis. Metamorphosis is most dramatic in the case of anurans, although there are significant changes that occur in caudates and, to a lesser extent, the caecilians. There are notable exceptions of neotony among the caudates. The most well-known and common of these neotenic species being the Mexican axolotl which retains its gills and larval characteristics throughout its lifetime. Metamorphosis is a complex process largely under the control of the endocrine system in larval amphibians. The main driving force of metamorphosis is the thyroid, producing varying levels of T3 (triiodothyronine) and T4 (thyroxine) to trigger and sustain the process.

Integument

Amphibian skin is naked and is permeable to semi-permeable depending on the species. Aside from forming a physical barrier to the environment, amphibian skin plays a role in osmoregulation, respiration, thermoregulation, and immune regulation. The outermost layer of the epidermis (stratum corneum) is only one cell layer thick. The underlying epidermal layer (stratum germinativium) is 4-8 cell layers thick depending on species. The dermis is also comprised of two layers- the stratum spongiosium (with the majority of chromatophores as well as the mucous and granular glands) and the underlying stratum compactum (compact cartilaginous fibers). In caudates and caecilians, the dermis is tightly adhered to the underlying tissue, with no functional subcutaneous space.

The permeability of the skin is dependent on the species and life history of the amphibian. In general, aquatic species have thinner, more permeable skin and allow for more free water exchange and cutaneous respiration. Species found in drier environments (toads, some species of frogs) have more keratinized or granular skin. While this reduces water loss in drier environments, it also can limit the respiration across the skin in these species. This may also have clinical consequences for transcutaneous drug administration. These species may rely on ventral pelvic drink patches for proper osmoregulation. These pelvic patches are hypervascularized areas of skin along the ventral pelvis in terrestrial species. In many species of caudates and anurans, ridges and small bumps along the skin may both increase the surface area for cutaneous respiration and aid in water absorption through capillary movement of water through these channels.

Numerous secretory glands are present within amphibian skin. These glands function to produce a mucus coat (mostly mucopolysaccharides) to maintain proper skin hydration in most species. In some species, these glands have become more specialized to produce varying toxins for defense. In toads, bufodienolides (cardiotoxic steroids) can be present throughout the skin. Some species (such as *Rhinella marina*) have well developed parotid glands over the dorsal cervical region that may shoot this toxin several feet as a defense mechanism. In the wild, dendrobatid frogs can produce highly toxic steroidal alkaloids through assimilating proteins in their diet. In captivity, these species do not produce these toxins due to their different prey selection. When threatened, some species of salamander are able to produce thickened mucoid secretions from the skin as a threat deterrent.

Gastrointestinal system

All adult amphibians are primarily insectivorous/carnivorous. In most species, prey capture is a highly visual process. In amphibians with reduced vision (either in murky water or a lack of eyes in the case of some cave salamanders), a sensory lateral line system many be used for prey detection. Olfaction likely plays a minimal role in prey detection and is used for more of a chemosensory role.

The alimentary system of adult amphibians is simple in structure, as a result of their carnivorous/insectivorous diet. Teeth are present in the majority of species. In many anurans and some salamanders (Plethodontids), a highly developed tongue and supporting hyobrachial structure are present for catching prey. At rest, the anuran tongue is flipped backwards in the buccal cavity and projected forward during the feeding response. The tongue is then retracted back into the mouth by the adductor musculature to ingest prey.

The esophagus is generally short and thin-walled with longitudinal folds to allow for expansion with prey items of various sizes. All species are monogastric with a simple stomach structure. The intestines are generally uniform in diameter without major morphological distinctions until the widened, thin-walled colon. The colon terminates into the cloaca. The liver is bilobed in anurans, and elongated in salamanders and caecilians. The gall bladder is present and empties directly into the duodenum. The pancreas is enclosed in the hepato-gastic ligament.

As larvae, there are various feeding strategies and associated alimentary changes across species. Larval anurans may be herbivorous, insectivorous, or filter feeders in the water. Larval anurans have keratinized mouth parts and elongated intestinal tracts to accommodate these feeding strategies. This more complex alimentary architecture and keratinized mouth parts are lost during metamorphosis.

Cardiovascular system

Adult anurans have a three chambered heart, with two atria and one ventricle. There is wide variation to the internal structure of the heart between taxonomic classes, and likely between species. Anurans have a complete atrial septum and internal septum, whereas the internal septum in caudates and caecilians is fenestrated. Both a renal portal system and hepatic portal system are present in amphibians.

The lymphatic system of amphibians is highly developed with extensive lymphatic tracts that vary based on taxonomic grouping. Lymph hearts beat independently of the cardiac rhythm to provide unidirectional flow to the lymph fluid. Caecilians have an extensive system of lymph hearts, numbering over 200 in some species. Anurans have fewer lymph hearts, but have developed extensive lymphatic sacs within the subcutaneous space. These sacs can be clinically used for medication and fluid administration. Lymph contains all of the components of peripheral blood, except for erythrocytes.

Respiratory strategies

Four main respiratory strategies are utilized in amphibians: Pulmonary, cutaneous, buccopharyngeal or branchial. Forms of respiration are not mutually exclusive and many species may rely on more than one form of respiration concurrently. Branchial respiration is employed by larval and neotenic amphibians. Respiration occurs across the capillaries of the gills in an aquatic environment. Branchial respiration is a passive process in neotenic species and larvae. Cutaneous respiration occurs passively across the skin either diffusely in some species or across thinner patches of skin in the case of some frogs and toads.

Buccopharygneal respiration is a characteristic of all known terrestrial amphibians. The nares are open to allow air to enter the buccopharyngeal cavity. In short bursts, the buccal floor is pumped up and down to move air in and out of the cavity rapidly. Gas is exchanged through the capillaries along the pharyngeal mucosa. In those species with lungs, intermittent pulmonary respiration will occur in conjunction with buccopharyngeal respiration. Pulmonary respiration occurs in the majority of anurans and many salamanders. Pulmonary respiration is an active process of a force pump mechanism. The amphibian lung lacks alveolar structures; gas is exchanged primarily through capillaries in the pulmonary tissue. Lungless salamanders (Plethodontids) rely on cutaneous respiration for 90% of their oxygen needs and have no functional pulmonary space.

Urogenital system

Kidneys in adult anurans and salamanders are paired, mesonephric, and highly vascular. Pronephric kidneys are present in larval amphibians, but are lost during metamorphosis. Kidney shape varies between amphibian species, with salamanders and caecilians having elongated paired kidneys. Kidneys in anurans tend to be shorter and wider. Anuran and caecilian kidneys are responsible for filtering peripheral blood as well as the coelomic cavity. Filtration primarily occurs at Bowman's capsules. Primary nitrogenous waste may be ammonia (in aquatic species), urea (most terrestrial species), or uric acid. In male salamanders and some anurans, the anterior portion of the kidney lacks functional glomeruli and tubules are used primarily for sperm transport. Wolffian ducts carry urine from the kidneys to the cloaca. A bladder is present in all known amphibian species and is ventrally connected to the cloaca.

In males, paired testicles are attached to the kidneys by the mesorchium. Testes may change in size depending on the seasonality and age of the individual. Anuran testes are generally spheroid, while in salamanders and caecilians the testes are lobed and elongate. In true toads, a portion of ovarian tissue develops at the anterior pole of the gonad (Bidder's organ). In females, the ovaries are suspended in the mesovarium along the mesial side of the kidneys. As ova develop, they are released into the coelom and moved into the oviduct through ciliated epithelium of the coelom.

Musculoskeletal system

The musculoskeletal system of amphibians varies widely by taxonomic group and ecological niche of the species. Caecilians are streamlined with no external limbs. They lack pelvic and pectoral girdles. Caudates typically have 4 limbs, which can be reduced in some species. Hindlimbs are reduced in *Amphiuma* and missing in *Siren* and *Pseudobranchus* spp. Most caudates are capable of limb regeneration following trauma or limb loss. Anurans have well developed hind limbs for saltatory movement for terrestrial species, and modifications for aquatic species. Most of the vertebrae are fused, and classified into presacral, sacral, and postsacral regions. The pelvic girdle is fused, and no sacrum is present. Many long bones are fused, including the radioulna, tibiofibular, and urostyle.

Husbandry

When evaluating the husbandry of amphibians, many of the same criteria are evaluated as in reptilian husbandry. However, some items become more critical to the care and success of amphibian management. Most importantly, proper attention to water quality becomes essential. All water for amphibians should be free of contaminants like chlorine and chloramines. This can be achieved with water conditioners or using properly aged or filtered water. Any standing water in amphibian enclosures should routinely be tested for pH, ammonia, nitrite, nitrates, and hardness. This becomes particularly essential in aquatics species like axolotls, African clawed frogs, or newts. Dissolved oxygen in the water is critical for cutaneous or branchial respiration, but is often difficult to accurately measure with at-home testing. Distilled or reverse osmosis water is not appropriate as a primary water source for amphibians due to the lack of essential minerals and solutes.

In general, compared to reptiles, most amphibians are maintained at cooler temperatures and are sensitive to heat stress and rapid dehydration. Both temperature and humidity should be constantly monitored to avoid even brief periods of temperature changes and inappropriate humidity. A thermal gradient should be provided to allow the animal to properly thermoregulate. Care should be taken to ensure the heating method for the enclosure does not rapidly dry the environment. Proper humidity levels can be maintained through periodic misting of the environment, automatic misting systems, moving water features, or a combination of methods depending on the needs of the species. When using bioactive setups, proper processing of plants and soil/leaves should be employed to prevent cross contamination of bacteria, fungus, or parasites.

Species specific recommendations are beyond the scope of this discussion. Where possible, consulting natural history resources of the species is warranted to determine the basic care of a species (temperatures, humidity, proper substrate, lighting, etc. The priority in properly housing amphibians should be to recreate a safe environment with proper temperatures, humidity, lighting, nutrition, and space to encourage natural behaviors for the species. In the case of commonly kept species, there are often guidelines for how to maintain them in captivity, although these should be properly evaluated by the clinician for their relevance and accuracy. ARAV care guides are available for some of the common amphibian species in captivity.

Handling

When handling any amphibian, care should be taken both to avoid injury to the handler and to the animal. For the handler, care should be taken to avoid bites from larger species (horned frogs, pixie frogs), as they are likely to bite and cause serious injury. Many toads also have well developed parotid glands over the back of the neck that may forcefully expel toxins when handled. Some salamanders are also capable of producing thickened mucous secretions in response to handling. Proper handling and personal protective equipment (gloves, with glasses or masks) are recommended. Some salamanders are capable of tail autonomy and will drop their tails with handling. Care should be taken to avoid putting any pressure on the tail. Before handling these species, the owner should be alerted to this potential autotomy.

Although there is mild risk of injury to the handler, it is much more likely for the handler to inadvertently harm or injure the amphibian during examination. Disposable gloves should be used for handling adult amphibians. Latex gloves should be avoided when handling larval anurans, as toxicity has been reported. Rinse gloves with distilled water immediately prior to handling. When working with multiple amphibians, a new pair of gloves should be used for each individual in an effort to reduce cross contamination of any infectious agents. Any oils, lotions, or soaps on the skin can be toxic to amphibians.

For all species, initial visual examination should be performed prior to a hands-on exam. Amphibians can be placed in a clear plastic container or clear plastic bag for visual exam. Use a clean container that has been properly disinfected and rinsed in between amphibian patients. Spray a small amount of distilled or filtered water in the container or bag to prevent dehydration during the exam. This containment allows for a full dorsal and ventral visual examination, reduces handling time, and can also be used as restraint for non-invasive diagnostics like radiographs and ultrasound.

Both the handling time and the number of times handled should be kept to a minimum to reduce stress on the patient. *Dendrobates tinctorius azureus* and *Lithobates pipens* exhibited elevated heart rates as a stress response to repeated handling, and this stress response should be assumed for all amphibian patients (Jacobson, et al. 2020).

For most species, room temperature should ideally be ~65-70°F (18-21°C) to prevent heat stress. Fresh distilled water should be readily available to rinse equipment and gloves, as well as to prevent dehydration.

Frogs and toads

Regardless of the size of the frog, standard handling techniques can be used for frogs and toads. Care should be taken to avoid any excessive pressure when handling, especially for smaller species like Dendrobatids. Frogs and toads can be safely gripped around the pectoral or pelvic girdle with just enough pressure to keep them secure. When handled in this way, all important structures should be visible and palpable (head, mouth, coelom, limbs), allowing access to mouth and limbs for medication administration and diagnostic sampling.

Salamanders and newts

Due to the slender body shape of most salamanders, physical restraint presents a major challenge. The majority of the examination should be visual prior to handling. When handling salamanders, the use of an appropriately sized clear plastic tube can be helpful to visualize the animal from all angles. When handling to collect samples, using wet soft towels or wet pieces of foam for positioning and gentle pressure can be helpful.

Amphibian examination

Initial visual examination in a clear escape proof container provides visualization of the majority of the epidermis as well as the eyes. Limbs can be observed for positioning and proper muscling. General posturing and movement can be observed. Respirations can be observed and recorded, depending on the species and primary mode of respiration. Nares should be clean and free of mucus/debris. Heart rate can be assessed via doppler through a plastic container or bag, at the level of the heart or the pelvic patch.

After a thorough visual exam, the animal can be picked up for a physical examination. Coelomic palpation should be performed on all species. Coelomic palpation may be complicated in some species due to defensive posturing and lung inflation. The coelom should overall be soft and compliant, but the clinician may appreciate food in the stomach from a recent meal (or foreign bodies). Transillumination may assist coelomic evaluation in some species. The mouth can be opened with a soft, thin speculum (soft guitar pick, laminated paper, business card or old credit card as appropriate based on the species). Care should be taken to carefully open the mouth with minimal pressure to avoid mandibular fractures. In many species, there is a small lip of the maxilla over the mandible which necessitates slipping the speculum under the maxilla and over the mandible to get the mouth open. While in hand, the limbs can be palpated and outstretched to check the skin along the limbs and the flanks that was not previously visible. In smaller anurans and caudates, the heart may be visible through the skin and a heart rate can be recorded.

During the examination, the clinician should constantly be monitoring the patient for signs of stress, including but not limited to thrashing, mouth gaping, signs of dyspnea or bleeding. If these signs are noted, the animal should be placed back in a safe, moist container for observation. With handling, some amphibians may vocalize a warning call or micturate when picked up.

Diagnostic sampling

Wound/ulcer evaluation

A common clinical presentation for amphibians is the presence of a skin ulceration or wound. All wounds should be checked with an impression smear or skin scrape for the presence of parasites or bacteria. Impression smears are made by lightly touching a clean glass slide to the lesion, picking up a thin layer of mucus and other surface contaminants. These slides can then be directly evaluated as a wet mount, or air dried and stained with Diff quik, Gram stain, or Acid fast stain for the presence of different pathogens. Treatment can then be targeted and initiated based on these results.

Skin lesions may be present with *Batrachochytrium dendrobatidis* (Bd, chytrid, chytridiomycosis) in some species (retained shed, erythema, change in skin texture), but the lack of skin lesions does not rule out Bd infection in any animal. Clinical signs of infection range from no signs in subclinical carrier to sudden death in more susceptible species. The mechanism of disease is through the disruption of the normal function of the skin (barrier to environment, as well as for osmoregulation). Tadpoles may be minimally or subclinically affected, as keratinocytes are only present along the mouth parts.

Fecal evaluation

Fecal evaluation is a common and useful tool in evaluating overall amphibian health. Both wet mounts and floatation of fecal samples should be performed. Samples can be collected directly from the amphibian's normal environment or ideally from a clean environment to reduce the chance of environmental contamination. Amphibians can be moved into a clean container with a wet paper towel as substrate to collect a clean sample for evaluation.

As with reptiles, amphibians may have low levels of parasites present in their fecal samples that could be considered insignificant or potentially normal flora. Fecal examination should always be assessed in regards to the clinical presentation for the animal. High levels of parasites or ova combined with the presence of red or white blood cells may be an indication of clinical disease. Proper treatment should be initiated based on the parasite present and clinical condition of the patient.

Nematodes are likely the most common and clinically significant parasites to be encountered by the clinician. Lungworms (*Rhabdias spp.*) and intestinal nematodes (*Strongyloides spp.*, among others) are common findings in captive and wild frogs. Due to their direct lifecycle, there is the potential for continued reinfection in a captive setting. For this reason, treatment of the animal and the environment are indicated for clinical infections. Routine fecal monitoring of previously infected individuals is recommended.

Blood collection

The general guideline for blood collection volume is 10% of blood volume in the clinically healthy patient. Standardized values for hematologic and biochemical values are limited in amphibians, but values for some species have been evaluated. Hematologic and biochemical values should be evaluated in terms of the overall clinical picture for the animal. Hemoparasites can also be identified in blood samples.

<u>Anurans</u>

There are multiple blood collection sites described in anurans. Site selection depends on multiple factors: 1) the size of the anuran, 2) clinical condition of the patient, and 3) comfort of the clinician in blood collection. In most anurans, blood may be collected from the lingual plexus, ventral abdominal vein, femoral vein, heart, or from the facial/musculocutaneous vein in Ranidae.

<u>Caudates</u>

In caudates, the ventral abdominal vein, femoral vein, and heart may be used for blood collection using methods similar to anurans. The ventral tail vein is an additional site of blood collection in caudates. Similar to reptiles, the vein courses along the ventral spine on midline. Caution should be used in collecting from the ventral tail vein in some salamanders as tail autonomy is possible. Sedation may lower the risk of tail autonomy.

Diagnostic imaging

Radiography and ultrasonography can be utilized in amphibians as in other species. In most cases, restraint for radiographs and ultrasound can be achieved by placing amphibians in a clean plastic bag moistened with filtered or distilled water. Radiography allows for the evaluation of the musculoskeletal system, as well as the GI tract and lungs (when present). Contrast media may assist in the evaluation of the GI tract. Ultrasonography can be used to evaluate cardiac rhythm, further evaluate organomegaly or edema/hydroceolom, and aid in aspiration of the coelomic cavity for diagnostic sampling or therapeutic drainage.

Necropsy

Described in detail elsewhere, necropsy of individuals is a valuable tool for the clinician to not only determine the cause of death in individuals of a collection, but to review the anatomy of both common and uncommon species. Ideal necropsy specimens are freshly deceased and removed from the enclosure as quickly as possible (ideally within the first few hours). Autolysis occurs rapidly in amphibians.

Therapeutics

Medication administration

Medication route is highly dependent on several factors which include: 1) the type of medication to be administered, 2) the species of amphibian, and 3) the desired effect of the medication. The main routes of medication administration are: intramuscular injection, oral administration, topical or bath solution (transcutaneous), environmental, subcutaneous/intralymphatic, intracoelomic or intravenous injections. At this time, there is a lack of clinical knowledge for the majority of medications across various species. Clinicians are encouraged to consult primary literature for more information on medication dosing.

Although doses are commonly extrapolated from one species to another, use caution when making assumptions across species and especially across taxonomic groups. A recent paper evaluated plasma concentrations of ceftazidime in Northern leopard frogs (*L. pipens*) via subcutaneous and transcutaneous administration. Plasma concentrations only remained above therapeutic threshold for 24 hours after administration of 20mg/kg ceftazidime SC, and were too low to determine in the transcutaneous group (Hawkins et al 2021). This is a shorter dosing frequency than the recommendations in some reptiles. In contrast, population pharmacokinetics suggested a ceftazidime dose of 20mg/kg subcutaneously every 5 days in easter hellbenders (Musgrave et al. 2021). More research is needed to make species specific recommendations for dosing of even common medications.

Intramuscular injection is a direct method of application that is advantageous in that it bypasses the need for medications to be absorbed through the GI tract or the skin in possibly debilitated species. It does require that the clinician has an appreciation for the muscular anatomy of amphibians, and that the amphibian is large enough for a safe injection. While the renal portal system is present in amphibians, it is reduced compared to reptiles. The hepatic portal system may affect drug absorption from the caudal half of the body.

The proper use of oral medications relies on the ability of the clinician to open the mouth of the amphibian for direct application, or to get the animal to eat a prey item containing the desired medication. This is highly dependent on the species of amphibian. Care should be taken in larger anurans and salamanders to avoid potentially harmful bites.

Topical or bath solutions are a common therapy for bacterial or fungal infections, especially those affecting the skin. Topical solutions can be administered at the appropriate volume over the dorsum of most species, or along the ventral drink patch for toads or amphibians with thickened skin. Due to the high vascularity and permeability of the skin, many of these medications are readily absorbed and, in some studies, have been detected in organs at effective concentrations. One example is the transcutaneous administration of enrofloxacin in Coqui frogs (Valitutto, et al. 2013). Caution should be used with novel medications, as some medications may cause cutaneous reactions or may not achieve sufficient systemic concentrations from topical application.

Environmental therapies can be employed for some diseases. In many cases with sick amphibians, cooler temperatures may provide a better alternative to temperature increases to prevent heat stress. *B. dendrobatidis* can be cleared with elevated temperatures for set periods of time. However, most amphibians cannot tolerate the higher temperatures needed for therapy.

Subcutaneous/intralymphatic injections can be used to administer fluids or medications directly into the lymphatic vasculature of anurans. The subcutaneous space is not available for therapeutic use in caudates and caecilians. There are a series of lymphatic sacs along the dorsum of most anurans. The most clinically accessible lymphatic sacs are the paired lymphatic sacs along the dorsum cranial to the pelvic bones, immediately cranial to the iliac crest. Injections into the lymph sacs could be used as an alternative to intravenous injections.

Intracoelomic injections have been described as a route of drug or fluid administration in some species. Intraceolomic injections of propofol and sterile solutions of tricaine methanesulfonate have been described in amphibians with varying dose ranges and effects. Intracoelomic injections can be administered in the paralumbar fossa or ventrally lateral to midline. Care should be taken to avoid damage to any internal organs. The effect of coelomic filtration on the kidneys may alter drug metabolism.

Intravenous administration of drugs is limited in amphibians. Most commonly, euthanasia solution can be injected intravenously or intracardiac in sedated animals for euthanasia. Intravenous drugs may have some value for chemotherapeutics or anesthesia, but these medications have not been evaluated in any studies. Sedation is recommended for intravenous drug administration in most species.

Anesthesia and analgesia

Although there is a lack of pharmacokinetic or pharmacodynamics studies for many medications used in amphibians, sedation and anesthesia has comparatively been well described in the literature. Like other medications, there is variability in the effect of each medication in different species. One should not expect the same dose of one medication to work in all species of amphibians, and more studies are needed to determine pharmacokinetics/dynamics in commonly encountered species. As with other poikilothermic animals, all medications are temperature dependent in amphibians. Amphibians should be maintained within their POTZ for optimal drug metabolism. The following discussion is not meant to be an exhaustive list of all anesthesia and analgesia options, but a review of the current literature. More specific recommendations on dosing can be found in current textbooks and peer-reviewed literature.

Tricaine methanesulfonate

Tricaine methanesulfonate (MS-222) has been well described for amphibian anesthesia, and varying dose ranges have been described. MS-222 works as a muscle relaxant by inhibiting action potentials across neurons. MS-222 comes as a white crystalline powder that readily dissolves in water. In freshwater, MS-222 needs to be buffered to prevent major pH decreases. The solution can be buffered with sodium bicarbonate at a volume equal to or double the weight of the dry MS-222 powder. MS-222 has been shown to cause retinal toxicity in humans with chronic exposure, so proper PPE is recommended when working with the dry powder. Once prepared in liquid solution, MS-222 will break down quickly when exposed to light.

Dose ranges from 0.5-2g/L have generally been used for amphibian anesthesia and sedation, although higher doses have been used in toads and for euthanasia. In general, aquatic species with less keratinized skin require lower doses for anesthesia compared to species with more keratinized skin (toads). Induction of anesthesia can be achieved by placing the amphibian in a plastic container with a shallow amount of MS-222 and monitoring for decreased movements and ultimately a lack of righting response. Once under anesthesia, the amphibian can be maintained under anesthesia by rinsing fresh MS-222 solution along the body. Depth of anesthesia can be determined by monitoring heart rate via Doppler, respiratory rate, and response to noxious stimuli. If an animal is too deep under anesthesia, they should be rinsed with fresh water until awake.

MS-222 alone was not sufficient for euthanasia in smokey jungle frogs (*Leptodactylus pentadactylus*), with spontaneous movements noted in frogs for prolonged periods of time after induction (Balko et al. 2019).

Eugenol/Isoeugenol

Eugenol is the active ingredient in clove oil. Eugenol has been used topically or by immersion to induce sedation or anesthesia in amphibians. In leopard frogs, deep anesthesia was induced by a 15-minute bath, but caused reversible gastric prolapse in 50% of frogs (Guenette and Lair 2006). In *Xenopus laevis*, eugenol via immersion results in a surgical plane of anesthesia for 30 minutes with respiratory depression but normal heart rate (Guénette, et al. 2007). In another study, topical eugenol resulted in cutaneous necrosis in *X laevis* (Ross, et al. 2006). Adverse effects have been noted in *Rhinella marina* with death, respiratory distress, and noncardiogenic pulmonary edema (Hernandez, et al. 2012). Isoeugenol has been used in *Litoria ewingii* tadpoles, resulting in dose dependent induction and recovery (Speare, et al. 2014).

Propofol

Propofol has been evaluated in several species via intravenous injection, intracoelomic injection, or by immersion. When administered intravenously to *Lithobates catesbeianus*, propofol did not induce a change in heart rate, and only caused an abbreviated increase mean arterial pressure (within 5 minutes) (Williams et al. 2018). As with other anesthetic agents, propofol's effects may be dependent on species. When administered at 35mg/kg intracoelomically, deep anesthesia was induced in tiger salamanders (*Amblystoma tigrinum*), but only sedation was noted in Sonoran Desert toads (*Bufo alvarius*) (Wojick, et al. 2010).

Alfaxalone

Alfaxalone is a neuroactive steroid used in veterinary medicine for induction of anesthesia. It works as a positive allosteric modulator on $GABA_A$ receptors and can act as a direct agonist of $GABA_A$ receptors. It is metabolized in the liver and has a short half-life in dogs and cats. Alfaxalone can generally be used as sedation or anesthesia for minor procedures, but provides no analgesia. For deeper anesthesia or more painful procedures, adjunct analgesia should be provided.

Alfaxalone increased the heart rate of bullfrogs (*L. catesbeianus*) after IV administration for approximately 45 minutes. There was a transient increase in mean arterial pressure with administration that significantly decreased to hypotension after 10 minutes (Williams et al. 2018). Tachycardia, but not hypotension, was noted with IM alfaxalone at the same dose.

Subcutaneous administration of alfaxalone in blue poison dart frogs (*D. tinctorius azuerus*) resulted in light sedation for 40 to 60 minutes with response to tactile stimulation. Cardiovascular depression and apnea were noted in 20 and 30mg/kg doses (Yaw et al 2020a).

Alfaxalone by intramuscular injection resulted in dose dependent sedation in Spanish ribbed newts (*Pleurodeles waltl*), with a high dose resulting in a reduced induction time and prolonged recovery (Cermakova et al 2020). Intramuscular alfaxalone in three species of *Litoria* frogs found variable responses to the same dose of alfalaxone Sedation and bradypenia was induced with 30mg/kg in *L. caerulea*, and with 20mg/kg in *L. aurea*. In *L. booloolongensis*, a light plane of anesthesia and bradypenia were induced at 20mg/kg (Sladakovic, et al 2014).

Alfaxalone sedation via immersion and transbrachial application has been reported for anesthesia in a Mexican axolotl (*Amblystoma mexicanum*) undergoing surgery for foreign body removal (McMillan and Leece 2011). Alfaxalone by immersion has been evaluated in fire bellied toads, *Bombina orientalis*. Sedation, but not surgical anesthesia, was achieved in 11 of 12 individuals at 200mg/L (Adami et al 2015). When combined with morphine, toads showed decreased nociception. No changes to nociception were noted with the addition of butorphanol by immersion (Adami et al 2016a). When combined with dexmedetomidine by immersion, additional analgesia was noted, but the depth of anesthesia appeared to be reduced compared to alfaxalone alone (Adami et al 2016b).

Inhalant agents

Sevoflurane and isoflurane have been studied in amphibians. Inhalant isoflurane has been described in mountain chicken frogs (*Leptodactylus fallax*) for the placement of intracoelomic radiotracers. Frogs were induced at 5% isoflurane, intubated and maintained on isoflurane at 2% for the procedure with no adverse effects (Barbon et al. 2019).

Topical isoflurane mixed with a water-based gel induced a loss of righting reflex in American tree frogs (*Hyla cinerea*), but resulted in erythematous skin lesions and signs of systemic illness following application. Topical sevoflurane mixed with a water-based gel in *H. cinera* induced a loss of righting reflex, with a recovery time 4.5 times faster than isoflurane with no adverse effects noted (Zec et al. 2014). A similar topical application of sevoflurane jelly in *R. marina* resulted in consistent loss of righting reflex, with a higher dose leading to longer recovery times (Stone et al. 2013).

Anecdotally, the use of isoflurane- or sevoflurane-soaked cotton balls in an enclosure induction chamber have been used for anesthetic induction in amphibians. This method likely results in increased levels of anesthetic gas than is provided by the vaporizer and results in rapid induction of anesthesia. To the author's knowledge, no peer-reviewed studies have evaluated this method of induction.

Combination protocols

A comparison of subcutaneous administration of alfaxalone-midazolam-dexmedetomidine versus ketamine-midazolam-dexmedetomidine in blue poison dart frogs (*D. tinctorius azuerus*) resulted in a rapid loss of righting reflex and similar recovery times of approximately 96 minutes. Increased pulmonary respiratory depression was noted with the alfaxalone combination, but the heart rate was similar between protocols (Yaw, et al. 2020b).

Nonsteroidal anti-inflammatories

Meloxicam at 0.1mg/kg has been shown to decrease circulating levels of prostaglandin E2 levels in North American bullfrogs (*L. catesbeiana*), suggesting an anti-inflammatory effect (Minter et al. 2011. However, in *X. laevis* at 0.2mg/kg injected into the dorsal lymph sac, no signs of anti-nociception were noted (Coble et al 2011). Smokey jungle frogs (*L. pentadactylus*) administered meloxicam 0.2mg/kg SC or topically did not achieve detectable plasma concentrations at any time point studied. In comparison, ketoprofen administered at 1mg/kg topically or SC resulted in detectable enantiomers for 3-24 hours with a decline over time (Balko, et al. 2018). In a separate study, administration of firocoxib SC, ketoprofen SC, meloxicam SC, or topical diclofenac did not result in clinical changes in nociception via acetic acid testing (Balko, et al. 2020).

Flunixin meglumine at 3.3mg/kg achieved plasma concentrations considered to be therapeutic in cattle after a single topical dose in 21 *R. marina*. There were no significant pathologic signs associated with medication use in a subset of 11 toads examined histologically, except for a possible superifical skin lesion in one individual (Scott, et al. 2020).

Flunixin meglumine has been evaluated in two studies in *X. laevis*. Flunixin meglumine at 25mg/kg injected into the dorsal lymph sac provided better analgesia via acetic acid test and Hargreaves testing compared to meloxicam, morphine, and xylazine. One frog in this study died following administration of flunixin meglumine, with bilateral kidney congestion and interstitial inflammatory infiltrates (Coble 2011). In another study, 25 and 50mg/kg flunixin meglumine was administered to *X. laevis* undergoing anesthesia for oocyte harvest. In both doses, mild to moderate myocyte degeneration and necrosis were found around the area of injection in the dorsal lymph sac. Worsened necrosis was found in animals sedated with benzocaine prior to administration, compared to MS-222. Death occurred in 5 of 12 individuals in the 50mg/kg dosing group (Smith, et al. 2018).

Opiods

The majority of opioid use and data in amphibians is experimental, with few clinical studies on efficacy. A review of pain and the use of opioid medications in amphibians has been published (Stevens 2011)

Buprenorphine has been evaluated at high doses experimentally in Eastern spotted newts (*Notophthalmus viridescens*). A return to normal behavior after administration of 50mg/kg intracoelomically was observed following limb amputation (Koeller 2009). This high dose of medication may be impractical for most clinical cases. Buprenorphine has not been shown to be effective in reptiles, and should be used with caution in amphibians.

Butorphanol by continuous immersion at 0.5mg/L for 3 days resulted in a return to normal behavior in Eastern spotted newts (*N. viridescens*) following forelimb amputation (Koeller 2009).

Disease processes

In-depth discussions of many amphibian diseases are covered elsewhere in textbooks and primary literature. Clinicians are strongly encouraged to consult relevant veterinary textbooks and the ever-expanding body of literature for more specific information on amphibian disease in both captive and wild populations. As with other species, a thorough history review, physical examination, and appropriate diagnostics should be used to develop an informed differential list and treatment plan.

Conclusion

Although amphibians are not commonly encountered in general practice, a thorough understanding of anatomy, physiology, husbandry and management can greatly inform diagnostic and treatment options. These are the building blocks used to properly diagnose and treat underlying disease. More research is needed for species specific recommendations of therapeutic medications.

Literature cited

1. Jacobson C, Doss GA, Yaw TJ, et al. Effect of manual restraint and visual security on heart rate in dyeing poison dart frogs (*Dendrobates tinctorius azureus*) and leopard frogs (*Lithobates pipiens*). J Herp Med Surg. 2020;31:59-63.

2. Hawkins SJ, Cox SK, Sladky KK. Pharmacokinetics of ceftazidime in Northern leopard frogs (*Lithobates pipiens)* at two different doses and administration routes. Am J Vet Res. 2021;82:560-5.

3. Musgrave KE, Hanley CS, Papich MG. Population pharmacokinetics of ceftazidime after a single subcutaneous injection and normal oral and cloacal bacterial flora survey in eastern hellbenders (*Cryptobranchus alleganiensis alleganiensis*). J Zoo Wildl Med. 2021;52:90-6.

4. Valitutto MT, Raphael BL, Calle PP, et al. Tissue concentrations of enrofloxacin and its metabolite ciprofloxacin after a single topical dose in the coqui frog (*Eleutherodactylus coqui*). J Herp Med Surg. 2013;23:69-73.
5. Balko JA, Posner LP, Chinnadurai SK. Immersion in tricaine methanesulfonate (ms-222) is not sufficient for euthanasia of smokey jungle frogs (*Leptodactylus pentadactylus*). J Zoo Wildl Med. 2019;50(1):89-95.

6. Guenette SA, Lair S. Anesthesia of the leopard frog, *Rana pipiens*: a comparative study between four different agents. J Herp Med Surg. 2006;16:38-44

7. Guénette SA, Hélie P, Beaudry F, et al. Eugenol for anesthesia of African clawed frogs (*Xenopus laevis*). Vet Anaesth Analg. 2007;34:164-170

8. Ross A, Guénette SA, Hélie P, et al. Case of cutaneous necrosis in African clawed frogs *Xenopus laevis* after the topical application of eugenol. Can Vet J. 2006;47:1115-1117

9. Hernández SE, Sernia C, Bradley AJ. The effect of three anaesthetic protocols on the stress response in cane toads (*Rhinella marina*). Vet Anaesth Analg. 2012;39:584-590

10. Speare R, Speare B, Muller R, et al. Anesthesia of tadpoles of the southern brown tree frog (*Litoria ewingii*) with isoeugenol (Aqui-S). J Zoo Wildl Med. 2014;45:492-6.

11. Williams CJ, Alstrup AK, Bertelsen MF, Jensen HM, Leite CA, Wang T. Cardiovascular effects of alfaxalone and propofol in the bullfrog, *Lithobates catesbeianus*. J Zoo Wildl Med. 2018;49:92-8.

12. Mitchell MA, Riggs SM, Singleton CB, et al. Evaluating the clinical and cardiopulmonary effects of clove oil and propofol in tiger salamanders (*Ambystoma tigrinum*). J Exot Pet Med. 2009;18:50-6.

13. Wojick KB, Langan JN, Mitchell MA. Evaluation of MS-222 (tricaine methanesulfonate) and propofol as anesthetic agents in Sonoran desert toads (*Bufo alvarius*). J Herp Med Surg. 2010;20:79-83.

14. Yaw TJ, Mans C, Martinelli L, et al. Evaluation of Subcutaneous Alfaxalone for Sedation in Juvenile Blue Poison Dart Frogs (*Dendrobates tinctorius azureus*). J Herp Med Surg. 2020;30:248-53.

15. Cermakova E, Oliveri M, Ceplecha V, et al. Anesthesia with intramuscular administration of alfaxalone in Spanish ribbed newt (*Pleurodeles waltl*). J Exot Pet Med. 2020;33:23-6.

16. Sladakovic I, Johnson RS, Vogelnest L. Evaluation of intramuscular alfaxalone in three Australian frog species (*Litoria caerulea, Litoria aurea, Litoria booroolongensis*). J Herp Med Surg. 2014;24:36-42.

17. McMillan MW, Leece EA. Immersion and branchial/transcutaneous irrigation anaesthesia with alfaxalone in a Mexican axolotl. Vet Anaesth Analg. 2011;38:619-623

18. Adami C, Spadavecchia C, Angeli G, et al. Alfaxalone anesthesia by immersion in oriental fire-bellied toads (*Bombina orientalis*). Vet Anaesth Analg. 2015;42:547-51.

19. Adami C, d'Ovidio D, Casoni D. Alfaxalone–butorphanol versus alfaxalone–morphine combination for immersion anaesthesia in oriental fire-bellied toads (*Bombina orientalis*). Lab anim. 2016;50:204-11.

20. Adami C, d'Ovidio D, Casoni D. Alfaxalone versus alfaxalone–dexmedetomidine anaesthesia by immersion in oriental fire-bellied toads (*Bombina orientalis*). Vet Anaesth Analg. 2016;43:326-32.

21. Barbon AR, Routh A, Lopez J. Inhalatory isoflurane anesthesia in mountain chicken frogs (*Leptodactylus fallax*). J Zoo Wildl Med. 2019;50:453-6.

22. Zec S, Clark-Price SC, Coleman DA, et al. Loss and Return of Righting Reflex in American Green Tree Frogs (*Hyla cinerea*) after Topical Application of Compounded Sevoflurane or Isoflurane Jelly: A Pilot Study. J Herp Med Surg. 2014;24:72-6.

23. Stone SM, Clark-Price SC, Boesch JM, et al. Evaluation of righting reflex in cane toads (*Bufo marinus*) after topical application of sevoflurane jelly. Am J Vet Res. 2013;74:823-7.

24. Yaw TJ, Mans C, Martinelli L, et al. Comparison of subcutaneous administration of alfaxalone–midazolam–dexmedetomidine with ketamine–midazolam–dexmedetomidine for chemical restraint in juvenile blue poison dart frogs (*Dendrobates tinctorius azureus*). J Zoo Wildl Med. 2020;50:868-73.

25. Minter LJ, Clarke EO, Gjeltema JL, et al. Effects of intramuscular meloxicam administration on prostaglandin E2 synthesis in the North American bullfrog (*Rana catesbeiana*). J Zoo Wildl Med. 2011:680-5.

26. Coble DJ, Taylor DK, Mook DM. Analgesic effects of meloxicam, morphine sulfate, flunixin meglumine, and xylazine hydrochloride in African-clawed frogs (*Xenopus laevis*). J Am Assoc Lab Anim Sci. 2011;50:355-60.

27. Balko JA, Watson MK, Papich MG, et al. Plasma concentrations of ketoprofen and meloxicam after subcutaneous and topical administration in the smoky jungle frog (*Leptodactylus pentadactylus*). J Herp Med Surg. 2018;28:89-92.

28. Balko JA, Posner LP, Ossiboff RJ, et al. Safety and Efficacy of Topical Diclofenac and Subcutaneous Firocoxib, Ketoprofen, and Meloxicam in Smoky Jungle Frogs (*Leptodactylus pentadactylus*). J Herp Med Surg. 2020;30:101-6.

29. Scott G, Louis MM, Balko JA, et al. Pharmacokinetics of Transdermal Flunixin Meglumine Following a Single Dose in Marine Toads (*Rhinella marina*). Vet Med Int. 2020.

30. Smith BD, Vail KJ, Carroll GL, et al. Comparison of etomidate, benzocaine, and MS222 anesthesia with and without subsequent flunixin meglumine analgesia in African clawed frogs (*Xenopus laevis*). J Am Assoc Lab Anim Sci. 2018;57:202-9.

31. Stevens CW. Analgesia in amphibians: preclinical studies and clinical applications. Vet Clin: Exot Anim Pract. 2011;14:33-44.

32. Koeller C.A. Comparison of buprenorphine and butorphanol analgesia in the eastern red spotted newt (*Notophthalmus viridescens*). J Am Assoc Lab Anim Sci. 2009;48:171–175.

Additional Resources

Baitchman EJ, Herman TA. Caudata (Urodela): Tailed Amphibians. In: Miller RE, Fowler ME, eds. *Zoo and Wildlife Medicine Vol 8*. St Louis: Saunders/Elsevier; 2015:13-20.

Chai N. Anurans. In: Miller RE, Fowler ME, eds. *Zoo and Wildlife Medicine Vol 8*. St Louis: Saunders/Elsevier; 2015:1-13.

Clayton LA, Mylniczenko ND. Caecilians. In: Miller RE, Fowler ME, eds. *Zoo and Wildlife Medicine Vol 8*. St Louis: Saunders/Elsevier; 2015:20-26.

Duellman WE, Trueb L, eds. *Biology of Amphibians*. Baltimore: Johns Hopkins University Press; 1994.

Mendez D, Webb R, Berger L, Speare R. Survival of the amphibian chytrid fungus *Batrachochytrium dendrobatidis* on bare hands and gloves: hygiene implications for amphibian handling. Dis Aquat Organ. 2008;82:97-104.

Pessier AP. Chytridiomycosis. In: Mader DR, Divers SJ, eds. *Current Therapy in Reptile Medicine and Surgery*. St Louis: Elsevier; 2014:255-270.

Pessier AP. Infectious Disease of Amphibians: It isn't Just Redleg Anymore. In: Mader DR, Divers SJ, eds. *Current Therapy in Reptile Medicine and Surgery*. St Louis: Elsevier; 2014:247-254.

Pessier AP, Mendelson JR, eds. A Manual for Control of Infectious Diseases in Amphibian Survival Assurance Colonies and Reintroduction Programs. Apple Valley, MN: IUCN/SSC Conservation Breeding Specialist Group. Available at: http://www.amphibianark.org/wp-content/uploads/2018/07/Disease-Manual-2017.pdf

Whitaker BR, McDermott C. Amphibians. In: Carpenter JW, ed. *Exotic Animal Formulary*. 5th ed. St Louis: Elsevier; 2017:53-80.

Wright KM. Amphibian Therapy. In: Mader DR, Divers SJ, eds. *Current Therapy in Reptile Medicine and Surgery*. St Louis: Elsevier; 2014:281-288.

Pasmans F, Martel A. Amphibian Taxonomy, Anatomy, and Physiology. In: Divers SJ and Stahl SJ, eds. *Mader's Reptile and Amphibian Medicine and Surgery*. 3rd ed. St Louis: Elsevier;2019.

Sladakovic I, Divers SJ. Amphibian Anesthesia. In: Divers SJ and Stahl SJ, eds. *Mader's Reptile and Amphibian Medicine and Surgery*. 3rd ed. St Louis: Elsevier;2019.

Whitaker BR, Wright KM. Amphibian Medicine. In: Divers SJ and Stahl SJ, eds. *Mader's Reptile and Amphibian Medicine and Surgery*. 3rd ed. St Louis: Elsevier;2019.

Yaw T, Clayton L. Differential Diagnoses by Clinical Signs- Amphibians. In: Divers SJ and Stahl SJ, eds. *Mader's Reptile and Amphibian Medicine and Surgery*. 3rd ed. St Louis: Elsevier;2019.

AUTHOR INDEX